# Sons of the Sea Goddess

ANTONIUS C. G. M. ROBBEN

# SONS OF THE SEA GODDESS
## Economic Practice and Discursive Conflict in Brazil

COLUMBIA UNIVERSITY PRESS
NEW YORK

**Columbia University Press**
New York • Oxford



**Library of Congress Cataloging-in-Publication Data**

Robben, Antonius C. G. M.
Sons of the sea goddess : economic practice and discursive conflict in Brazil / Antonius C.G.M. Robben.
p. cm.
Bibliography: p.
Includes index.
ISBN 0-231-06842-5.—ISBN 0-231-06843-3 (pbk.)
1. Maritime anthropology—Brazil—Camurim. 2. Fishers—Brazil—Camurim. 3. Economic anthropology—Brazil—Camurim. 4. Camurim (Brazil)—Economic conditions. I. Title.
GN564.B6R63 1989
306.3—dc20 89-31704
CIP

Casebound editions of Columbia University Press books are Smyth-sewn and printed on permanent and durable acid-free paper

*Book design by Ken Venezio*

*To my parents Antonius J. Robben*
*and Andriëtta Robben-Ackerman*

# Contents

# Preface

This book arises out of a dissatisfaction with economic anthropology's inattention to the interpretations of economic practice made by the economic actors themselves, and from the firm conviction that instead of neglecting these interpretations they should become the basis of our ethnographic accounts. Unlike ethnographers in other branches of anthropology who in the last decade have successfully deepened their understanding by paying close attention to what people say about their own culture, most economic anthropologists still seem to regard economic actors as people who can only provide them with factual data but whose interpretations cannot be trusted, and who only in rare instances have at best a very dim awareness of the economy they are reproducing through their daily actions, decisions, and strategies. They believe, following in the footsteps of Malinowski (1961:83), that people experience only part of the economy in which they are involved, that they only know their own motives and objectives, and that therefore they cannot see the total picture of the institutions and structures which guide their economic behavior. This may be so, yet, it is through the tension among these partial views that the actors constitute the economy. Our failure to respect this diversity of fragmentary visions in our analyses, either by dismissing those interpretations as false or by treating economic actions as guided by shared folk models, assumes a uniform constitution of the economy that neither exists in the practices nor in the discourse of the actors.

Many economic anthropologists assume, furthermore, that most people do not have a sufficient grasp of the modern world system and the global market economy to offer insightful interpretations about how their operation affects them. It is true in the case of the fishing economy examined in this book that the Brazilian fishermen have very little knowledge of the complexities of the postindustrial society that influences their lives in unsuspected ways. The fishermen of Brazil do not have the same awareness of the distant horizons of Western culture as the anthropologist does, and we cannot demand from them the interpretive acumen and verbal adroitness of some privileged informants of tribal societies who have blessed their ethnographers with penetrating interpretations of their own culture. It is within the bounds of their limited knowledge and the plainness of their words that the Brazilian fishermen understand the world. We cannot expect from them a reflective sophistication, so characteristic of our Western world, which they have not cultivated. However, people do not only interpret the world through speech and discourse but also in practice, in the decisions they take, the acts they carry out or refuse to carry out, the emotions that move them, and the way in which they lead their lives or are forced to lead their lives. And here the fisherman's mastery greatly surpasses that of the anthropologist. Poised on evermoving canoes and balancing rapidly across the narrow boards of a small boat, the Brazilian fishermen are reproducing the economy about which they reflect only when the flow of practice breaks down in dispute and disagreement. It is around these two dimensions of interpretation—discourse and practice—that this ethnography of the Brazilian fishermen of Camurim has been written.

My research among Brazilian fishermen dates back to 1977 and 1978 when I spent close to one year in two coastal communities in the state of Alagoas in Northeast Brazil. The decline of raft fishing and the emergence of boat fishing in one of the villages stimulated me to do the present study on the relationship of two different fishing modes. I decided upon the community of Camurim after a month long survey of over a dozen towns in the states of Bahia and Rio de Janeiro. Camurim had a vital pluriform fishing economy composed of canoe fishermen and boat fishermen, urban wholesalers who were purchasing fish for the large metropolitan centers in the south, a stratified social structure with great differences in wealth and power among the social classes, and a budding tourist industry which had not yet reached the proportions of the more accessible beaches along the Bahian littoral. Ironically, Camurim is less than a day's sailing away from the place where Brazil was discovered yet is situated in the country's least-developed region along the Atlantic Ocean, with tropical rainforests that at my arrival still loomed tall and imperious over the dirt road leading to Camurim. I spent most of the time between August 1982 and December 1983 in Camurim, but left on several trips to Salvador and Rio de Janeiro to collect material from archives, libraries, and government agencies. Since then, I have made one visit to Camurim in August 1987, and learned that the boat fishing and canoe fishing modes were just as stable and vigorous as in 1982–83.

My research in Camurim was made possible by a generous grant from the National Science Foundation (BNS 8200732). I thank Elizabeth Colson and Eugene Hammel for their detailed comments on my research proposal. John Cordell kindly invited me to participate as a researcher in his project for World Wildlife Fund on the conservation and indigenous management of coastal habitats in Brazil. His dedication to the study of inshore territoriality inspired me to look for other forms of sea tenure in Camurim. A fellowship from the Michigan Society of Fellows at the University of Michigan in Ann Arbor allowed me to thoroughly rewrite this book. The challenging intellectual environment of the Society with its eclectic gathering of members strengthened me in my search for knowledge beyond the boundaries of anthropology.

I am greatly indebted to Burton Benedict, who skillfully guided me through my graduate studies at the University of California at Berkeley. He has the rare quality of being able to enter into an author's reasoning and criticize the work from within without imposing his own thoughts. I owe him much for his encouragement in writing this book.

A number of other persons deserve credit. Gerald Berreman, Stanley Brandes, and Bernard Nietschmann contributed valuable suggestions to the doctoral dissertation out of which this book has been written. Bob Scholte at the University of Amsterdam played the role of devil's advocate. His scholarship and stimulating queries obliged me to examine many theoretical questions in more depth and explore areas of whose existence I was unaware. It fills me with great sadness that death did not allow him to see this book before it went to press.

I want to thank my friend Smadar Lavie for the years of companionship in Berkeley, both intellectual and personal, and for the many inspiring dialogues, some of which kept us on the phone till midnight. James Anderson of the University of California has always freely shared his insights on maritime cultures with me. Dwight Heath of Brown University gave valuable comments on chapter 8. I want to give special thanks to Luiz Fernando Dias Duarte of the Museu Nacional in Rio de Janeiro. Our common interest in fishing cultures brought us together more than a decade ago, and his erudition and continued friendship have helped my research in numerous ways. I thank the library staff of the Museu Nacional for their assistence. Hélio Ranssen of the SUDEPE in Salvador and Joel Pereira da Costa, the president of the Fishermen's Guild in Niterói, taught me much about fishing. Their knowledge and assistance were indispensable. I am indebted to Luz Marina Ruiz for expertly printing the photographs for this book, and to Rachael Cohen for wordprocessing the various drafts of this manuscript. I am grateful to Pma Johnstone, Carolyn Nordström, Marcelo Suárez-Orozco, and Marcela and George Nazzari for their friendship and encouragement.

I am thankful to Raymond Kelly, Conrad Kottak, Bruce Mannheim, Sherry Ortner, and James Boyd White at the University of Michigan for the many ideas they let loose on me. I also treasure the stimulating discussions at the biweekly faculty seminar of the Program in the Comparative Study of Social Transformations which sharpened my argument in many ways. I owe a special debt to Conrad and Betty Kottak, Martin and Shirley Norton, and Rodrigo Díaz-Pérez for welcoming my wife and me to Ann Arbor.

My warmest thanks go to Célio and Maria Angela Assunção, who received us into their home as if we were members of their family. Their hospitality made us feel Brazil in our hearts. We cannot imagine what our experience in Camurim would have been like without their haven in Rio de Janeiro. I also cherish the long conversations about Brazil with my dear friend Lucas Assunção, wherever they were held, on the slopes of Corcovado, the beaches of San Conrado, or at the Guanabara.

I owe my greatest debt to the people of Camurim. They readily accepted my wife and me into their lives, opened their homes to us, and were always willing to answer our puzzling questions about the obvious. We remember their friendship and miss their presence every day.

My research and this book could not have been done without the dedication and love of my wife, Claudia. She not only made the illustrations but also made me describe those experiences which she thought necessary to understand Camurim. This book is as much hers as mine.

I have a lasting debt to my parents. They have always supported my studies relentlessly, even though they knew that their encouragement would inevitably make us live apart for many years to come. I dedicate this book to them.

**Sons of the Sea Goddess**

# 1
# Introduction:
# The Demystification of Economy

Valdemar placed his offering with care on the waves. Fishing had been poor the last few weeks and his youngest daughter had fallen ill. He needed a good catch urgently. The flowers, the shawl, and the turtle shell comb bobbed up and down, sometimes turning over, and then drifting off with the current. Each abrupt movement of the offering filled Valdemar with anticipation, only to disappoint him again with its reappearance on the surface. If the present remained afloat, then the Sea Goddess had refused it; if it went down, it had been accepted. The distance between the gift and the boat became greater until only a glimpse could be seen. The fishermen assumed their usual positions on deck. Valdemar laughed nervously, shrugging off the ritual as pure entertainment for the crew. Suddenly, he felt a strong tug at his fishing line. Iemanjá, he mused, had forgiven him after all for his meager present. He began to brag about his luck and how he was going to make the catch of his life, when a hollow piece of black coral with a flat base appeared. The captain saw in the coral a vase sent by Iemanjá as a sign of anger at Valdemar's show of disrespect. The Sea Goddess allows fishermen to catch fish for their sustenance but She will only help those who reciprocate her kindness with honor and dignity. Valdemar caught only nine kilograms of fish during the ten-day trip and stayed ashore for more than a week afterwards, convinced that the supernatural wrath of the Sea Goddess would prevent him from catching any fish at all.

There is a strange paradox in the way anthropologists conceive of the relation between economy and society. We all believe in the existence of an economy as a domain of society and can somehow differentiate economic relations and actions from other social practices, but we fail to trace the boundaries and establish the precise location of the economy in society and culture. Clearly, what we consider as economic and noneconomic is more determined by the conventions of our discipline and the empirical issues we have addressed traditionally in our ethnographies than by unambiguous epistemological demarcations. Yet it is on this belief in a boundary between economy and society that economic anthropology has constructed its paradigms. The economy is treated as a bounded social system despite the lip service paid to the embeddedness of the economy in society.[1] The cost of Valdemar's offering will certainly be considered economic but his belief in the supernatural power of the Sea Goddess will not be incorporated into the models or principles that intend to explain the fishing economy.

This book wants to challenge the persistent myth that the economy is a bounded domain of society and culture that therefore can be truthfully represented in models, structures, laws, and principles. I will argue that the conflicting interpretations among our informants about the place of economy in society and culture can neither be ethnographically ignored, nor theoretically reasoned away by assuming that ultimately everybody in a particular culture makes economic decisions in the same manner because of an underlying folk model, a universal drive towards maximization, or an all-embracing form of social integration or opposition. To do empirical and analytical justice to the diversity of opinions about the economy that arise from different social positions of power and interest, the multiple discourse demands a hermeneutic pluralism because each interpretation is based on different premises about social reality.[2] With this study of a Brazilian fishing economy I want to argue against the positivist reductionism prevalent in economic anthropology, and demonstrate that we have to make the conflicting interpretations of economy part of our ethnographic descriptions and theoretical constructions because it is through these discursive conflicts that economic practice is formed and transformed.

This ethnography of the Brazilian town of Camurim describes how boat owners, boat fishermen, and canoe fishermen have succeeded in developing a viable alternative to the unrewarding plantation work that dominates the regional economy. A pluriform fishing economy has emerged in which a canoe fishing mode and a boat fishing mode coexist, each characterized by its own recruiting practices, fishing grounds, catching methods, channels of distribution, fish dealers, markets, and price setting conventions. However, each principal economic group has a different conception of the fishing economy. Boat owners regard themselves as subject to an institutional conglomerate of banks, federal agencies, and large fish marketing corporations whose business decisions are guided by the impersonal forces of supply and demand. Boat fishermen, instead, perceive the economy as part of an interactional universe with social relations that transcend the institutional boundaries perceived by the boat owners. The boat fishermen believe that the economy operates through such qualities as friendship, graft, favoritism, and manipulation which are not restricted to economic relationships but which interlace the institutional links perceived by boat owners with other social ties. The canoe fishermen share this interpersonal conception of economy with the boat fishermen but add that boat fishing is an expansionist mode of production driven by a desire for capital accumulation and political power. Boat owners and boat fishermen attempt to appropriate the traditional fishing grounds of the canoe fishing mode and entice canoe fishermen with the ease of motorized fishing and the promise of boat ownership.

These conceptions of the fishing economy as institutional, interactional, and expansionist are not just conflicting interpretations of the same economic relations and practices, but these contradictions become part of economic practice as multiple structural complexes. In addition, each group encapsulates different aspects of society into its economic universe, and makes economic decisions on these different grounds. What may be purely religious for one group may be considered entirely inside the economic realm for another group. Yet the existence of economic boundaries is acknowledged by all Camurimenses. Whenever I use the term "economy" in this book I will, therefore, refer to this shared belief in such a social domain of society, even though the fishermen disagree on where those boundaries should be drawn. Canoe fishermen relate their productive practices closely to their domestic world. They work to raise the standard and quality of living of their families, and fear that the expansionism of the boat fishing mode will take away their independence as producers and heads of household. Boat owners and boat fishermen, instead, attribute little direct significance to the household but focus on the public world with its extensive social networks, displays of masculinity, and pursuit of prestige, power, and influence. This intertwinement of the economy with other social spheres is further substantiated by the behavior of fishermen who switch between canoe fishing and boat fishing. Former boat fishermen become much more interested in the household after they move to canoe fishing, while former canoe fishermen develop public aspirations. The ethnographic analysis of family life and the public world of bars and brothels in Camurim becomes therefore an integral part of this study of the local fishing economy.

Given the different conceptions of the pluriform fishing economy held by the fishermen of Camurim, how is the economy at all possible? How can the fishermen cooperate without misunderstanding each other's aims and actions? How can a myriad of actions ever add up to a functioning social institution when the actors disagree on whether or not those actions pertain to the economic domain? I will demonstrate throughout this book that the economy is constituted through a dialectic process of which practice and discourse are two different dimensions. Discursive conflict and routine practice are not mutually exclusive but can coexist because they pertain to different modes of interpretation. People can disagree about the meaning of fishing when considered in a wider economic context, but they may still agree on how a particular goal can be successfully achieved or on how the many actions that make up that practice should be carried out, for instance, by joining hands in raising a sail or jointly killing a large shark. The economy is constituted through practice and reconstituted through its dialectic with discourse. Conflicting discourse may temporarily suspend economic practice, reflect on its social consequences, reconsider its direction, and influence the conditions of its resumption. This dialectic process from practice to discourse and back to practice lies behind the dynamic of economic change and continuity.

Discursive conflicts in Camurim do not arise only about the economy but also about people's world view. What struck me most during my first weeks in Camurim were the many disagreements about the causes of natural events. I often joined a group of canoe fishermen resting on the beach under the late afternoon sun as they watched the waves rolling towards them. Our vision enclosed by the curved horizon, I asked about the earth, the winds, and the tides. Several men said that the earth was a flat disc, but a few young fishermen believed it to be round. Another group assured me that the earth had the shape of a bowl with the ocean at its center. They attributed spring tides to the excessive tilting of the earth-bowl. Others assumed that periodically the moon absorbs and releases large amounts of water. At spring tide, they reasoned, the yellow color of the full moon becomes visible because all water has flown back to earth, while the dark, new moon signals the absorption of water at neap tide. The more fishermen I talked to, the greater the variation became. The fishermen of Camurim did not have a shared interpretation of the natural world, but had a practical wisdom that allowed them to deal effectively with natural phenomena such as the relationship between tides and the phases of the moon. I could have used Western scientific criteria to decide whether their explanations were correct or not, but this knowledge would have been irrelevant for them as long as their practical understanding was sufficient to establish the tidal rhythm.

When I turned my questions to the social world, and to the economy in particular, I encountered a comparable array of opinions except that the arguments were more passionate. But here

I could not distinguish as easily between right and wrong, because I was placed in a preinterpreted social world. I did not deal with a natural, physical universe of storms, currents, and tides that could never impose its meaning upon me, but I was confronted with people who offered me conflicting interpretations of their actions, interests, motives, and history. Since these interpretations guided their economic practices, I could not rely on theoretical criteria to differentiate between true and false. I could only reach a situated understanding that placed their statements in context.[3] Instead of searching for underlying structures or objective processes that would supersede these conflicting perspectives or reconcile them on a higher plane of abstraction, I wanted to understand how the economy was constituted in contested practice and discourse. Did these discursive conflicts arise from the economic practices? How did these contradictory conceptions of economy become part of economic practice? How could fishermen cooperate despite this dissonance? What was the effect of this divergent economic understanding on other aspects of their lives? Did the fishermen reach a deeper understanding of their social world through conflict? Could the tension between practice and discourse lead to change? This book is about these problems of interpretation, understanding and conflict among the fishermen of Camurim, about their diverging goals in life, and about communication in practice under the weight of fundamental dissension.

## Practice and Discourse

More than any other branch of sociocultural anthropology, economic anthropology has tried to follow a pre-paradigmatic route to normal science (Kuhn 1962). Its relentless search for basic principles, structures that distinguish the essential from the contingent, and models that can make reliable projections of economic trends, has yielded a number of schools of thought which make equally strong claims to truth.[4] The quantifiable character of economic data and the success of economics as the most rigorous of the social sciences have contributed substantially to the pursuit of this positivist ideal. If we assume that economic decisions are techno-rational and that economic phenomena can be priced, measured, weighed, and counted, then there is no need to complicate our analyses with the redundant interpretations of the economic actors. Only improved research methods and more sophisticated analytical tools will give economic anthropology the scientific maturity so desired.

Despite this optimism, I believe that theoretical disputes waged in isolation of the actors impoverish our understanding of economic practice. I do not want to diminish the importance of economic models for the purpose of designing policies, or for comparative and theoretical uses, but I emphasize that such models cannot anticipate or construe the contextual interpretive process through which economic continuity and change are forged. Only more knowledge of this constitutive process will allow us to improve our understanding, because the economy is as much suffused with interpretation and meaning as any other realm of culture, such as religion. Both religion and economy construct the human condition through practice and discourse, even though the economy takes a material form that shrouds its interpretive nature.

My interpretive approach has been influenced by the practice or structuration theorists Bourdieu, Giddens, Foucault, Heidegger, Marcuse, and Sahlins.[5] Practice theory postulates the primacy of practice over reflection and cognition. Culture exists first and foremost in what people do, not just in what they say and think about their routine or in the cognitive models that are often believed to pattern their actions (Bourdieu 1977:2; Giddens 1979:39; Ortner 1984:149–150).[6] Instead of analyzing symbols, rituals, social systems, deep structures, or historical processes, practice theory focuses on ways of coping in the world, on practice as people's interpretation of defining themselves socially and culturally. The most crucial and defining quality of people is to be found not in what they think but in what they do and how they do what they do, namely, attending to the everyday things, tasks, and demands of their social world.[7]

Practices are the rock bottom of humanness and therefore cannot be reduced to underlying beliefs, laws, rules, scenarios, or cognitive processes.[8] The argument behind this radical standpoint is that people do not need explanations and guidelines when the practices are performed successfully. The canoe fisherman who masters an action—baiting a fishhook, making a sacrificial offer to Iemanjá, reciting a spell to impair a colleague from catching fish—does not need to interpret its meaning consciously in order to complete the task. He has a practical wisdom which demonstrates a situational understanding of his activities (Gadamer 1985:274–289; Hoy 1978:55–61). Practical wisdom is "tacit knowledge that is skillfully applied in the enactment of courses of conduct, but which the actor is not able to formulate discursively" (Giddens 1979:57; cf. also Giddens 1984:6–9). This is not to suggest that practices are either performed automatically or at random. I am arguing, together with Polanyi (1958), that the regulation of practice does not depend merely on recipes of proven sequential action but is greatly influenced by a sense of personal judgement developed through hands-on experience which, for instance, allows a fisherman to maintain the right tension in a fishing line to secure a fish. Such practical wisdom can never be expressed in rules or principles but only acquired in practice. Personal skill combined with a cursory agreement on the goal and execution of a particular activity allow fishermen to coordinate their actions without sharing the same interpretation of the practice of which those actions are part.

Some practices into which Camurimenses are socialized at an early age, such as the appropriate personal space between a fisherman and his patron, are so routine that they are not questioned. People learn how to act appropriately as they acquire a sense of the social world through an interpretive process that is abandoned as soon as they have attained mastery (Cicourel 1970). A child is told to show respect to a landowner by maintaining a proper distance, but this maxim cannot be grasped or enacted without the practical knowledge of actual situations and experiences (cf. Polanyi 1958:30–31, 49–57). The child learns through practice, not through explanation. Hence, the interpretation of this routine practice by an actor is not necessarily more insightful than that of the anthropologist, as both are unaware of the dialectic of discourse and practice that preceded it.[9] What distinguishes actors from anthropologists is that the interpretations of the actors have a practical relevance which may lead them to alter their actions.[10] These interpretations need, therefore, to become the basis of our ethnographic accounts.

Habitual practices like the personal space between people seem to exist as objective ways of coping in the world (Berger and Luckmann 1966), and have become part of people's habitus or array of commonsense, taken-for-granted social ways (Bourdieu 1977:72, 1984:170). According to Bourdieu, Giddens, and Sahlins, these habitual practices may eventually develop structural properties which may escape the actors. These structures or "schemes are able to pass from practice to practice without going through discourse or consciousness" (Bourdieu 1977:87). These structures are neither causal nor teleological, and do not exist independent from the actors who reproduce them. They give continuity to social practice by providing a medium for the purposive, but not necessarily conscious, actions of social actors. The importance of structuration or practice theory is that it reconciles agency and structure into one approach. It demonstrates that the structural properties of practice "are both the medium and the outcome of the practices they recursively organize" in the process of structuration (Giddens 1984:25).[11] These structures are simultaneously produced and reproduced, interiorized and exteriorized as the unintended consequences of both conscious and habitual pursuits (Certeau 1984:57; Giddens 1984:8–14; Sahlins 1985:ix). However, despite their apparent social facticity, the structures of habitus and social practice are not as unequivocal as Bourdieu and Giddens suggest. The contradictory interpretations of economy become embodied in the practices themselves. The principal relations of the institutional, interactional, and expansionist conceptions of economy acquire a structural and enduring quality by being systematically singled out by the economic actors. The discursive contradictions become part of economic practice. Unintentionally, the boat owners, boat fishermen, and canoe fishermen constitute and reproduce those structures which they perceive to be characteristic of the economy. They perpetuate the contradictions of their discourse in the economy itself.

Economic practices are not only not guided by objective structures, but people also disagree discursively on what those structures are because of the following reasons. One, the terms of the structuration of practice are not identical for all actors. The child of a landowner who is socialized into the propriety of a spatial distance between social superiors and inferiors acquires a different structuration of face-to-face interaction than does the child of a sharecropper. The different sense of violation implied in a hierarchical relation between two persons is a clear indication. The powerful can transgress established conventions without provoking the repercussions which would be unleashed against those of lower status. Two, people may not have the same conception of the forms of structuration. The boat owners of Camurim find the economic structure of the fishing economy primarily in its institutional relations, while the boat fishermen perceive a more flexible economic universe of shifting alliances and social ties structured by vaguely defined expressions of friendship, favoritism, honor, and prestige. Three, people may disagree about the range of practices that are influenced by a particular structure because of differences in social position, power, or goals in life. Boat owners define economic practice much more narrowly than their crews, a definition which excludes religious as well as political concerns, while canoe fishermen pay much attention to the domestic world. Four, traditional practices have to be reproduced through actions, and here there is always the possibility that established customs are reinterpreted and challenged, or that old ways cannot respond to new contexts. This historical dimension makes the transformation of routine practice possible because "every reproduction of culture is an alteration, insofar as in action, the categories by which a present world is orchestrated pick up some novel empirical content" (Sahlins 1985:144). New conditions surrounding the routine practices may raise the structuring categories to the level of consciousness.

Practices sedimented in habitus are never sufficient to direct all courses of action in the social world or to cope routinely with the complex flux of everyday life. The fishermen of Camurim have to make sense of their own actions and decide about their direction in relation to the actions of others. In this process of interpretation they may get into conflict about what the practices mean, how they are structured, how they should proceed, or whether their actions fall outside the economic realm. The flow of action is halted and the actors enter into a mode of discursive interpretation which arises when the practice breaks down. These breakdowns of practice may either be partial or total.[12]

A partial breakdown helps to adjust people's actions to the context. The discourse evaluates their practical understanding of the social and natural environment, and tunes the actions without fundamentally challenging them. Partial breakdowns are common occurrences in fishing. For instance, two fishermen who fail to catch fish will deliberate about the use of different bait. Even though the flow of action is briefly interrupted, the question is not considered in a detached mood but is discussed within the immediacy of catching fish. This partial breakdown will reveal certain aspects of the context—the lunar cycle, sea currents, migratory movements—that are relevant to the activity at hand. Once the problem has been diagnosed and a solution formulated, the men will cast their fishing lines in the water and continue to fish.

A total breakdown happens when there are such serious misgivings about the practices that they are lifted out of their immediate context and discursively recontextualized in a wider frame of understanding which often involves the search for some inherent structure believed to pattern people's practices. For example, boat owners and their crews have repeated disagreements about their obligations towards the families of men who are at sea. Is the boat owner responsible for the welfare of a household in case of illness or should the fisherman make arrangements with his relatives in advance to take care of his wife and children? This issue cannot be easily resolved but dissolves into interpretations of the wider social context of the economic relationship. The fisherman argues that the capital owner has a duty as the employer to take care of his family, while the boat owner counters that the fisherman is self-employed and that his relatives are morally obliged to help his wife. There are no strict rules or norms to solve these discursive conflicts because each party places the dispute in a different context and interprets the economic relationship differently. The fisherman believes that the production unit exists to reproduce the basic needs of the family. The survival of the household is beyond all question. The boat owner says that he cannot guarantee the subsistence of his crew but that his enterprise merely provides an opportunity for the fishermen to turn those material conditions to their household's advantage. These discordant interpretations about the proper boundaries between the economic and the domestic world will continue to exist as part of the economic practice even though the contingent crises are temporarily resolved and the routine is resumed.

The outcome of this transition from discourse to practice depends on the balance of power among the actors in dispute. A skilled fisherman may force a boat owner into offering social assistance with the threat of resignation, while the boat owner may refuse to help a fisherman who is in financial debt. In both cases, there is a face-to-face confrontation in which the conflict becomes subject to the relative power of the two parties, and in which the interpretation of the more powerful is enforced into practice even though privately the subordinate party may continue to disagree.

Notwithstanding the persistent challenges to social inequality that contribute to discursive conflict, concealed shadow discourse may at times succumb to dominant discourse. Gramsci (1971:257–264) has referred to this ideological domination of shadow discourse as hegemony, while Bourdieu (1977:72) uses the term "habitus" in a similar way (cf. also Scott 1985:323; Terdiman 1985:59).[13] Bakhtin found the subordination of competing discourse in social "forces working towards concrete verbal and ideological unification and centralization, which develop in vital connection with the processes of sociopolitical and cultural centralization" (1981:271). Still, it is not just differences in power and social position which impel people to force multiple discourse towards ideological unification. It is also the hierarchical relations within social practices which provide space for the realization of taken-for-granted authoritarian discourse (Hirschkop 1986). The concept of hegemony adds a political dimension to discourse by drawing attention to the importance of practices of power, control, and dominance that may inculcate a reigning ideological stance through socialization. As I will describe in chapters 5 and 8, many former canoe fishermen who have switched permanently to boat fishing have been co-opted into a process of social mobility that legitimizes the social order and reinforces the power of the boat owners and the landed elite. They are socialized into practices of interpersonal submission which sometimes yield a higher social status but which most often entail only empty promises and economic deprivation for their dependents. By imitating the careers of boat owners, ambitious boat fishermen are unknowingly encapsulated into an ideology that presupposes hierarchy and subordination, and conceals inequality within the practice of social mobility. As Michael Polanyi (1958:53) has written:

> To learn by example is to submit to authority. You follow your master because you trust his manner of doing things even when you cannot analyse and account in detail for its effectiveness. By watching the master and emulating his efforts in the presence of his example, the apprentice unconsciously picks up the rules of the art, including those which are not explicitly known to the master himself. These hidden rules can be assimilated only by a person who surrenders himself to that extent uncritically to the imitation of another.

Michel Foucault (1978:92–97, 1982:208–226) has demonstrated how these latent force relations are reproduced in the routine micropractices of everyday life.[14] Foucault differentiates this "capillary" power from the "patrimonial" power vested in individuals, groups, social strata, and classes (Dallmayr 1984:84–89). Capillary power relations do not exist between subjects but among their actions. This power is embedded in what people do, even though they do not exercise it consciously but implicitly (Foucault 1982:220–224; Dreyfus and Rabinow 1982:187–197; Giddens 1979:88).[15] For example, there are many instances in which a captain-owner gives direct orders to his crew. He uses his legitimate power as the commander and owner of the boat to coordinate the actions of his men. Yet, his authority is further strengthened by many social and spatial arrangements that go beyond his personal power. He lives in a different neighborhood of Camurim than do the fishermen, he sleeps at sea in the motorcabin instead of below deck with his crew, his generosity in bars surpasses that of others, and his prolonged affairs with prostitutes are proofs of his masculinity. There is in these actions no intentional exertion of power based on the contractual hierarchy between boat owners and fishermen, but these practices reproduce and reaffirm deep relations of domination and subordination which appear in a more direct form in the fishing economy.

Formidable as the embedded force relations may be, they can never drown dissonant voices into complete complacency because people will always challenge the forces which restrict their participation in the social constitution of their lives. This opposition may not take the form of revolutionary action but hegemony is seldom so complete as to preclude dissent in hidden discourse and subtle acts of rebellion and sabotage (Giddens 1979:72; Scott 1985:319–340). These forms of defiance can never be entirely previsioned, anticipated, or preempted by a pervasive ideology because the context of dominance evolves continuously in new directions. Polemic discourse is as much a quality of social interaction as is hegemony (Bakhtin 1981:271–275; Todorov 1984:70). Paradoxically, the resistance to a dominant ideology may arise from the very force relations that are reproduced in the micropractices. The alternative interpretation of economy put forward by the boat fishermen of Camurim emerges from an awareness of their subordinate position in society. For example, their denunciation of a conspiracy between boat owners and wholesalers to suppress the price of fish may not be accurate or based on any hard evidence but comes from a deep-seated suspicion of all ideological justifications made by those in power, and this mistrust will give credence to even the wildest speculations. These confrontations will bring the contradictory structuration of economic practice to the surface of discourse. The practice of social inequality becomes open to dispute, and fundamental conflicts of interest in society are played out, bringing about social and economic transformations or instead touching off conservative attempts to contain the social forces of change. The combined study of concealed and dominant discourse will reveal the deep force relations and structures which the fishermen of Camurim perceive in routine economic practice (Foucault 1982:209–211; Taylor 1986:93–99).[16]

The fishermen of Camurim become aware of the contradictory structuring of their practices when the social routine breaks down, and they begin to search discursively for explanations and solutions. Confrontations send shockwaves through the practice, parts of which are revealed in discourse as structures. What was submerged in practice as unambiguous becomes ambiguous by its appearance through an acute awareness of habitual ways and mores. However, there is no consensus about the nature of those structures, because practices are not propelled by structures but by people's social position, interests, goals, and routines shaped through culture and context, and formulated in and influenced by discursive interpretations. The boat owners, boat fishermen, and canoe fishermen recognize the continuity of the pluriform fishing economy respectively in its enduring institutional, interactional, and expansionist structure, even though there are no objective grounds on which to validate either one of them. Nonetheless, economic anthropology has tried to find objective underlying structures or "universal models" (Gudeman 1986:28) valid for all economic actors in a particular society without realizing that various social groups may adhere to different terms and forms of structuration, identify different practices as reproducing those structures, and perceive a different context to their structured practices. The study of discursive conflict is therefore an important causeway into economic practice, because these confrontations occur at the junctures of people's conception of economy where economic contradiction, change and continuity are forged.

## *The Demystification of Economy*

The epistemological presupposition that economic models can describe the quintessence of economic action—although admittedly never replace it in all its empirical facets—is a positivist myth that has hindered our understanding of the interpretive constitution of economy. It has led to a pursuit of models, structures, and principles of economy, while taking the attention away from practical and discursive interpretation as the process by which actions take shape and get substance.

Demystification reveals the myth as myth and shows that people are at its origin (Ricoeur 1974:335–353, 381–401). The economy is not a system of objective relations which can be reduced to a few fundamental principles and axioms but is a whole of meaningful practices forged through confrontation and contestation. The discursive conflicts among the fishermen of Camurim help to strip the economy of its veil of objectivity because they demonstrate that economic practices cannot be separated clearly from other social practices. Disputes about the definition of basic needs, the morality of the catch division, the relation of capital and labor, and the operation of the market imply that there is no cognitive, folk, or cultural economic model held by all the participants of the fishing economy. This diverse intertwining of social and economic practices becomes most apparent in discursive confrontations which bear upon the cultural foci that figure prominently in the lives of the fishermen of Camurim.

A cultural focus is a center of practical orientation such as the Trobriand yam house and Kula valuables (Malinowski 1961), the Berber house (Bourdieu 1973), the French prison (Foucault 1977), the nucleus of a Spanish town (Gilmore 1977), and the Costa Rican market place and Church square (Richardson 1982).[17] The potlatch, cargo cult, Kula exchange, and Balinese cockfight are not cultural foci, because, unlike the above, they do not orient practices but are cultural practices themselves.[18] Cultural foci are material objectifications of cultural practices. They orient practices in material ways, just as the separation of the sleeping quarters of captain and crew directs the social interaction aboard ship. Yet cultural foci do not determine practices because people may disagree about the meaning of the habituses expressed in the cultural foci. The material form of a cultural focus can be hegemonic—as will become clear in my discussion of the spatial organization of the house in Camurim in chapters 6 and 7—but its interpretation and appropriation in practice are not.

The most significant cultural foci for capital owners and fishermen in Camurim are the boat and the canoe. A brief account of the development of Camurim's fishing industry demonstrates how the introduction of boats oriented the community in ways that profoundly changed the fishermen's conception of economy and society. The first boats plying the waters of Camurim were treated as if they were canoes. Crews went on short trips, visited familiar fishing grounds, and used the same fishing techniques as the canoe fishermen did. Once catches declined and costs rose, the captain-owners were obliged to operate the boats differently. They chose longer trips, at more distant locations, while using improved fishing equipment. None of these options was open to the canoe fishermen because of the material limitations of the canoe itself. This technical differentiation led to a greater social inequality between boat owners and fishermen, affected the conjugal relationships of the boat fishermen who spent increasingly more time away from home, and reoriented their economic objectives from raising the household's standard of living to the pursuit of capital ownership. The boat and the canoe were no longer technical implements but became cultural foci that reoriented the social practices of the actors of the pluriform fishing economy and represented, through their opposition, the principal differences between canoe fishing and boat fishing. Even though the Camurimenses decided about the introduction of boats and were thus the producers of the implements they began to use, this innovation entailed a much greater social and cultural importance than could have been foreseen because of an unknown technical and economic potential made possible by the material makeup of the motorboats.[19]

What sets boats and canoes apart from other technical means is their cultural centrality. They not only focus the economic practices, but also evoke discursive interpretations and lead to experiences that educe the links between economy and society. The wider significance of the boat and the canoe is sometimes mentioned openly among a group of men relaxing on the beach, but often it takes the shape of subtle experiences beyond the realm of production. The following excerpt from my field notes describes the grievances of a canoe fisherman who had just made his second trip on a boat. He experiences a fundamental power relation with the boat owner which is commonly verbalized with terms such as dependence (dependência) and submission (subordinação, sujeição, humildade).

> Watching Cabeludo and his son leave in a canoe to visit their nets, Lucas came strolling along the beach. He was dressed in shorts and a grey t-shirt with thin blue stripes. His eyes were slightly closed because he was still suffering a bit from conjunctivitis. "O, holandés? Como vai?" ["O, Dutchman? How are you doing?"], he exclaimed, "Tudo bem. E Você? Pegou muito peixe?" [Very well. And You? Did you catch much fish?"] He began telling how he had caught quite a lot of fish but that he was not thinking of making another trip because the expenses of the boat were very high. "Esse barco dá só pro Jacaré. Ele ganha até mais do que o pai." ["This boat gives only profit to (the captain) Jacaré. He earns even more than his father (the owner)."] He complained that he had been trying to find Jacaré to get his money. Yesterday he finally found the *mestre* of the boat. He was drinking with some friends. Jacaré complained that every time the boat arrived ashore, Lucas disappeared. Lucas answered that if he needed him, he could find him at home. Lucas does not want to spend all his hard-earned money on beer and rum. "Vou descer." ["I'm leaving."], said Lucas. He was going to João's bar to see if he could find Jacaré to get his account of the catch settled. I waited to see Cabeludo's catch. Nothing. I quickened my pace and caught up with Lucas.
>
> [Present in the bar annex store were João the bar owner, his wife, and three fishermen also waiting for their money.] The men were quiet, while the woman rattled on about ghosts, spirits, the 12:00 o'clock bus, the wind, and many other unrelated subjects. Out of respect for João's wife they nodded occasionally. All stared out of the door towards the late-afternoon movements of the town. The town's gardeners and sweepers were slowly trickling into the street beside the bar. They were tired. There were still five minutes

to kill. They sat on the sidewalk chatting, putting their implements in order. João's wife chattered on. Lucas got up. He looked tired and felt betrayed by the captain. The endless waiting for the money he had earned was killing him. He had seen Jacaré briefly when he arrived at the bar but the *mestre* had left again, probably to gamble with cards or to go drinking with his friends. Lucas and I strolled to the house of Cabeludo.

As this excerpt shows, the dependence of boat fishermen on captains and boat owners is not some ideological awareness of class that describes the abstract relationship of labor and capital, but it is emotionally experienced through the peer pressure to drink, the dishonest calculation of operating costs, the hide-and-seek power play of the captain, the hours loafing around the bar, and the putting up with idle chatter. Although these experiences would generally not be regarded as economic by most economic anthropologists or, as a matter of fact, by the boat owners, they nonetheless heap on the practice of boat fishing unavoidable corollaries of economic subjugation. The experiences of the boat owner are, of course, quite different. For him, the boat allows for opportunities towards status improvement and political influence, but also entails the dependence on a good fishing crew and the submission to the powerful wholesalers who control the marketing channels. Both the fisherman and the owner perceive the centrality of the boat in their economic practice but they experience, imagine and interpret that centrality differently. The economy is not a clearly demarcated domain of society and culture but a different whole of practices for different social groups.

Boat owners and boat fishermen can cooperate despite their discursive disagreements because they share an interdependence in the tasks to be carried out as well as a deeper dependence embedded in the micro-practices. This relation between practice and discourse is apparent from the contrasting interpretations in which three crew members holding Christian, African, and secular visions of the world expressed their relation to nature.

When Agenor caught very few fish during the normally profitable winter season and moreover lost several nets, he believed that he was under a spell and visited the spiritual leader of his Candomblé congregation to identify the evildoer. The helmsman, however, who was a staunch Catholic and a tireless foe of the Afro-Brazilian religion, told him that God must have had some higher reason for these hardships and that the only thing to do was to pray and have faith. The captain-owner, instead, excused the loss as bad luck and told Agenor to navigate to other fishing grounds. The three proposed strategies of action—manipulation, faith, and chance—did not immediately impede the cooperation of the men at sea. Yet, when the catches remained disappointing and in effect collapsed the fishing practice, Agenor was strengthened in his belief in the spell and left the crew while the others continued to test their faith and their luck.

When fishermen and capital owners talk about the economy, evaluate its place in society, and consider decisions that reach beyond the operation of the fishing unit, they perceive economic practices rather than an economic organization of actions and relations. Their thoughts do not stop at the functional boundaries of the fishing mode but encapsulate wider ramifications like Lucas' experience with boat fishing. For Lucas, economic practice is everything he considers relevant to fishing, including his fear of the high seas during stormy weather, dreams and fantasies about miraculous catches, mythologies about the Sea Goddess, and jokes about his masculinity after a weeklong separation from his wife. Lucas never compares boat fishing and canoe fishing in strictly monetary terms but he approaches them as encompassing wholes of practices.

This intermeshing of social and economic experiences is the basis for most heated controversies in Camurim. Fishermen are accused by boat owners of making unreasonable and irrational demands by confounding totally unrelated issues, such as the moral and contractual obligations of capital owners towards the families of their crews. The decomposition of cultural practice into separate social domains is therefore consciously advanced by boat owners and other wealthy townspeople who want to deal with fishermen on a narrow economic basis and refuse to assume larger social responsibilities. The ruling class benefits from treating socioeconomic issues as technical and depicting the economy as a bounded system of functional relations. Boat owners who can make fishermen believe that they organize their fishing enterprises exclusively on the principle of profit can exercise a much greater control over their crews. The demystification of the fishing economy of Camurim as a whole of practices transcending its alleged organizational boundaries shows, however, that this economic discourse of boat owners disguises personal interests and political ambitions.

Despite the strong orienting force of boats and canoes in Camurim, not everybody embraces these cultural foci as the exaltation of the economic world. One canoe owner's main interest lies in Candomblé, the Afro-Brazilian religion popular in Northeast Brazil. Zé Carlos guides his life from a religious perspective and follows ritual observances at sea that his colleagues regard as too elaborate. His cultural focus is the sacred center of his congregation. His canoe is little more than a fishing implement, and economic interests are subordinate to his religious concerns. Most fishermen fish on any day of the year, including Sunday and religious holidays. Only bad weather will prevent them from fishing or, as the saying goes, "The holy day of the fisherman is a southern wind." Yet Zé Carlos will sacrifice a good day of fishing and remove his nets to observe a religious festival.

The boat and the canoe are not the only cultural foci for the fishermen in Camurim. For the majority of the fishermen the house and the bar are the other important cultural foci in their lives. The bar is the cultural focus of public life, the house that of the domestic world. These cultural foci are not equally meaningful for all men. A man's preference for canoe fishing demonstrates, of course, why a canoe has a greater cultural centrality for him than does a boat. Likewise, the men do not attribute equal importance to the bar and the house. Canoe owners and canoe fishermen hardly ever visit the numerous bars of Camurim but spend most of their time ashore around the house. Boat owners and boat fishermen consider the house as little more than a conjugal obligation and prefer to stay in a bar with their colleagues and friends. At the risk of reification, one might say that the orienting forces of the cultural foci tend towards loosely consistent clusters of practice around which men organize their lives, and which contribute to the interlinkage of economic and social practices.

The fishermen themselves lean towards such reification when they talk about their friendship with these highly significant objectivations of cultural meaning. The following words of one canoe owner are typical: "After you sell or move away from a house, you might miss that house very much. You have made friends with the house. The same is true of a canoe. When a canoe is safe and fast, or you caught a lot of fish with that canoe, then you develop a friendship with her." Boat owners may have similar feelings for a boat or a bar. A certain bar may be regarded as a "home" (*casa de família*), the only place where the men feel entirely at ease. Boat owners have an especially great affection for their first vessel. They prefer to spend large sums of money on reconstructing their first boat than substitute a new one for it. One boat owner, who had to sell his first vessel to pay for the costly medical treatment of his autistic daughter, felt such an emotional loss that he constructed a similar boat which he appropriately named *Sufoco* (Suffocation).

The greater interest of canoe owners and canoe fishermen in the domestic world has to do with their more intimate association with the family (cf. Berger and Kellner 1964). They share actively the many small but highly significant affairs of the household and feel a commitment that is absent among the boat fishermen who have lost their control over this intimate social domain as they sojourn for a week or more at sea, only to return ashore for a few days filled with extended visits to bars and brothels.[20] Boat owners and boat fishermen regard the domestic realm as a domain of household relations without the strong emotional and symbolic meanings attributed to it by canoe owners and canoe fishermen. Their public and economic interests take precedence over the domestic concerns. They are mainly interested in the pursuit of a higher socioeconomic status, larger social networks, and greater economic and social mobility. Although at this point I cannot go into the many nuances which exist among the large group of boat fishermen, there is a general tendency to attribute more immediate importance to the public and economic world than to the domestic domain.

This analysis of the different clusters of cultural foci and practices considered by the fishermen of Camurim as central to their lives demystifies the belief in a bounded pluriform fishing economy which can be abstracted into a logically consistent model which subsumes the behavior of all economic actors. Economic practices are so intimately and so diversely related to other social practices that their postulated separation into demarcated social domains ignores the divergent interpretations of the fishermen of Camurim about the location of the fishing economy in society and culture.

## *Methodological Considerations*

The subtitle of this book refers to an emphasis on the practice and discourse of the fishing economy of Camurim as experienced and interpreted by fishermen and capital owners who are engaged in ongoing discussions that punctuate their daily routine. Open interpretational conflicts were abundant between boat owners and boat fishermen, but occurred only seldom between men who belonged to different fishing modes. In order to elicit such statements, I had to act as a relay of interpretations by confronting one with the statements of the other. These protracted dialogues took almost the form of dispatches in which I not only acted as the courier but also as the imaginary correspondent. For this reason, I do not present this study as a dialogue between ethnographer and informant—which is not to say that I discount my role as an intermediary who contributed to the dialogue—or as a conversation among informants whose self-reflective queries may have been prompted by my presence (cf. Clifford 1983; Rabinow 1977). Although I had to negotiate an understanding with my informants, the methodological goal was to elicit conflicting interpretations of practice.[21]

My use of the term "cultural focus" shows my influence on the presentation of the interpretations. Although boats, canoes, houses and bars were often topics of conversation, were considered formative forces in people's lives, figured centrally in the routine practices of everyday life in Camurim, evoked strong emotions, and appeared in dreams and fantasies, I have not only organized this ethnography around them but I also gave the cultural foci extra attention in my fieldwork. I would casually introduce questions about them into conversations among persons with either conflicting or consonant interpretations. Thus, I helped to orient the dialogues towards issues about which I needed more clarification but which were nevertheless of central importance to the people of Camurim. Rather than entering into discussions entirely of my own choosing, I tried to be guided by the topics that emerged in the dialogues among the men. As soon as I had written the arguments in a more or less crystallized form, I would return to the informants and ask for any necessary corrections, without of course jeopardizing those whose revelations were given to me in trust.

This ethnographic account contains my rendition of these dialogues and correspondences. I could have chosen a more dialogical mode of presentation, such as those of Crapanzano (1980) and Dwyer (1982), but that would focus the attention on the negotiation of an understanding between the informant and myself or, in case of a transcript of conversations among informants, omit those issues that cannot appear in face-to-face encounters because of the power difference between the actors (cf. Fowler 1985). A too narrow focus on public conflict between capital owners and fishermen would ignore "entirely the insinuations beneath the surface, the discussion outside the context of power relations" (Scott 1985:321). The objectification of the Other in ethnographic writing that Dwyer (1979) and Fabian (1983) have drawn attention to also exists within society itself.[22] All relations of dominance in Camurim—wholesalers and boat owners, boat owners and crew members, canoe fishermen and fish peddlers, landowners and plantation workers, men and women, parents and children—are characterized by an objectifying discourse that drives most fundamental dissent underground. Certain opinions are not for everybody's ear and can only be confided to an outsider. The anthropologist can be an outsider who has furthermore the privilege to incorporate such discordant opinions into seemingly spontaneous questions to opposite parties in order to elicit conflicting responses. For, what better way is there to understand an interpretation of the social world than through the most pertinent and penetrating questions posed by those who belong to the same society? Correspondence among anonymous parties via the anthropologist as their courier enables them to express their thoughts on matters they will avoid discussing in public, exactly because these issues touch upon the crucial relations of their social world.

Another hindrance to the uninhibited voicing of discontent and disagreement is the strong machismo of Brazilian fishermen. This attitude of male superiority often interferes in their conversations in public. Most heated discussions end up in verbal duels in which the importance of winning an argument supersedes the reasons for the dispute itself. The argumentation among groups of equally minded persons differs therefore substantially from the reasoning among antagonistic groups, either because the dissenting voices are silenced or because the assertions are inflated.

How can an ethnographer compare the relative validity of the conflicting interpretations when he or she does not accept the authority of formal, independent criteria? What constitutes evidence or what determines at least a better analysis? While one might judge the quality of informants in terms of the coherence, complexity, breadth, depth, and contextualization of their analyses, taken as collective interpretations the perspective of one social group is as valid as that of another. I tend to side with Rorty's hermeneutic critique (1979:315–322) that anthropology will never be able to discover a deeper unity or common ground among the conflicting interpretations of its informants and that therefore the various voices in the conversation should be allowed to show themselves. Although fishermen and owners have conflicting viewpoints, these do not stand in isolation. The interpretations assume a whole together with their negations. This indissoluble interconnectedness implies that there is not one right and several wrong interpretations—because they owe their existence to one another. The interpretations form a collage of views of the world that together are more complete than any perspective in particular.

This relativist position applies equally well to my ethnographic analysis of routine practice. My understanding of undisputed practice in a larger historical and cultural context is not inferior to that of a fisherman who before my queries never questioned the existence of certain customs and traditions himself. The principal reason is, as I stated earlier, that basic practices cannot be reduced to a limited number of decontextualized rules and variables. For instance, when I observed a canoe builder working on a dugout and asked him to explain what he was doing, he gave me a general summary of the sequential procedure and showed how he would solve the specific problems faced by this particular tree. Yet, even if he would explain his craftsmanship in increasingly more detail, I would never be able to make a dugout canoe without observing him and assisting him for months on end. Bourdieu has described this dilemma well:

> Invited by the anthropologist's questioning to effect a reflexive and quasi-theoretical return on to his own practice, the best-informed informant produces a *discourse which compounds two opposing systems of lacunae.* Insofar as it is a *discourse of familiarity,* it leaves unsaid all that goes without saying. . . . Insofar as it is an *outsider-oriented discourse* it tends to exclude all direct reference to particular cases. (Bourdieu 1977:18)

In other words, the informant has no explanation for the most obvious other than that this is the way of tradition. When, in turn, he or she realizes the practical incompetence of the anthropologist, he or she will superficially gloss over the intricacies of everyday living and verbalize a set of rules that must give the initiate some guidance. This repertoire of rules has very little practical significance because it has been decontextualized. The fisherman from Camurim has a practical wisdom on how to act appropriately but, when it comes to explaining these practices, his interpretation although more practically relevant is not necessarily better than mine. No judgment can be made on who is right or wrong, but one can only place one interpretation alongside the other, hoping that together they will raise our understanding of life in Camurim.

An alternative to this multivocal approach would be to construct a grand model that would incorporate the constitutive force of discursive conflict. Whatever context the actors perceive to their economic actions—whether supernatural, moral, ideological, or ecological—could be made part of this grand model. The major problem of this formalization is that it detaches a history of practice from its process of reproduction. The model is constructed in terms of abstract, decontextualized actions which can never keep up with the flow of everyday life. Not only practices change, but also their interpretations. What is valid today, is irrelevant tomorrow and not true the day after.

Another alternative might be to isolate the areas of consensus and use these seemingly hard facts as the bedrock of truth from which the rest can be judged. Consensus is easy to find. People often agree on the description of the catch division, the division of labor, the income scale, and even the enacted obligations and rights of capital owners and fishermen. They agree on the form and the order of events but they disagree on the context and the moral standing of these conventions. The fundamental misconception of this approach is that it denies conflict and discordance. If one were to concentrate on the conventions, then culture would be represented in its most shallow form and ignore that the areas of consensus are situated in larger spheres of fundamental disagreement (Taylor 1979:40). Conflict is a quality of society that cannot be reasoned away, because it enters the heart of the social construct. There will always be individuals or groups who will make others hear their dissenting interpretations. They give a new twist to the gamut of motivation and social behavior, and modify other interpretations and practices in their wake.

Understanding and knowledge are not static and absolute, but dynamic and historical. What seems the most insightful assessment at one moment may prove to be less than perfect and complete in another context or at another time. The understanding of the pluriform fishing economy is always in flux because the social world of Camurim does not have an immanent meaning—its meaning is created in the process of interpretation through talk and everyday living. Hence, it becomes impossible to discard certain interpretations as false and to deny the verity of the whole.

This conception of understanding and knowledge seems tautological. If understanding and knowledge become all that exists, then how can an anthropologist ever communicate his or her findings? What should be reported and what can be left out? I believe that the anthropologist should take his or her cues from the people themselves instead of claiming some privileged and superior outside knowledge on the basis of theoretical models or principles. The social groups on which I focused my attention presented me with a handful of interpretations. Once the relevant social groups identified themselves to me, I tried to provide an ethnographic context to their interpretations so that the conflicts became intelligible. The ethnographic context differs, of course, from the contexts in which fishermen situate their interpretations. My context tries to accommodate all relevant meanings, while the context perceived by a fisherman works like a sieve that retains only compatible meanings and creates uniformity instead of heterogeneity (Ricoeur 1974:71). The conflicts of these meanings and interpretations determine the crossroads at which the world is constructed. Here is where the brittle consensus that seems so obvious on the surface of social interaction breaks down. These conflicts must therefore form the clusters of meaning on which the ethnographic context should be focused.

The term crossroads is an apt metaphor that carries much significance for the fishermen of Camurim. When people want to overcome obstacles, be cured of illnesses, or in general improve their wellbeing, they can make a ritual offering (*despacho*) to Exú-of-the-Seven-Crossroads (Exú-das-Sete-Encruzilhadas). Exú is the messenger between the people and the gods (*Orixás*), who is often mistaken for the devil by those uninitiated in Candomblé religion. He is an Afro-Brazilian Hermes who resides at crossroads in forests, fields, villages, towns, and at all hidden, occult, and dangerous places in the world (Bastide 1978:251–255; Carneiro 1978:68–70). Just as requests for personal betterment are made at these intersections of the natural and the supernatural, so the arguments among fishermen and capital owners occur at the crossroads of meaning at which the economy is forged. Interpretation is inevitable in both cases, which shows that people's constitution and understanding of the social world is at heart hermeneutic, proceeding in circular rather than in linear ways (Gadamer 1985:259; Hoy 1978). Fishermen reflect through their discussions on the connection between their existence and everyday actions at home, at work, and in public. What is obscure in each of these social worlds acquires meaning by taking the perspective of the whole. This perspective takes a discursive form that reinvigorates the significance of the daily activities in a circular process of understanding. The boat, the canoe, the house, and the bar are cultural foci that shape and are shaped by what the fishermen and capital owners do. Ultimately, these cultural foci make them pose questions about their social and cultural existence, of what it is like to be a fisherman in Camurim.

# 2
# Ecology, History, and Community

"Exú the devil was waiting for Paiva do Otranto as he approached the crossroads at midnight," so began the canoe fisherman Manoel his tale of the rise to prominence of one of Camurim's most powerful men. "Paiva do Otranto did not ask for permission to pass but he walked right up to the Beast (*o Bicho*). In exchange for his soul, his blood, or his child," Manoel did not know what Paiva do Otranto had sacrificed, "he quickly amassed a fortune and became one of the richest landowners of Camurim."[1] Alberto, who was present at this conversation, gave me another version of the sudden wealth of Paiva do Otranto. Otranto did not strike a deal with the devil but had become a Freemason (*maçom*). As a member of the Masonic brotherhood, Otranto had received many profitable business tips from his brothers in faith. A couple of days after this conversation, I asked a boat fisherman about these rumors. He laughed and assured me that if there existed something like devil contracts, then half the town would gather at midnight at the crossroads to make a deal. He agreed with one of Otranto's employees, with whom I had talked several months earlier, that Paiva do Otranto had made his fortune by paying his workers miserable wages and making them work long hours. In the opinion of most fishermen in Camurim, social inequality and differences in wealth and power have been the result of

> ruthless practices, sheer exploitation, and an ambition unbridled by moral values.

Camurim is a town of nearly 6,000 inhabitants in southern Bahia located about 500 kilometers south of the state capital of Salvador and along the large cocoa-producing region of Brazil. The lives of the fishermen of Camurim are set against the background of a plantation economy with feudal overtones, a dominant ideology that legitimizes social inequality, and a community that measures status in terms of affluence and autonomy. The historical type of the colonial plantation owner who directed the lives of slaves, family, and clergy with almost absolute power is very deeply ingrained in Bahian society (Schwartz 1985:245–254), and became reinvigorated in the 1930s during the emergence of cocoa plantations in the county of Camurim. Manual labor became as degrading as land ownership became ennobling. A rendition of the colonial plantation system in terms of the dualism of master and servant, the ideological significance of land ownership, and the sociopolitical importance of the extended family is a simple yet essentially correct analysis (Iutaka 1971:257–260). What has come to constitute prestige in Camurim is the ability to live off one's possessions with a minimal involvement in their everyday operation. Land gives its owner a moral authority that translates into political power and high social status.

The poor Camurimenses who described the career of the landowner Paiva do Otranto disagreed about the path of mobility responsible for his rapid ascendence but they did not question the social hierarchy in which he assumed a prominent place. The sudden rise to power by unscrupulous individuals is a recurring theme in the history of Northeast Brazil; it is a proven model considered by many to be the only successful avenue of upward mobility. It is legitimized by a practice and ideology of inequality which have pervaded the social interaction and historical understanding of most social groups in Camurim. This sense of hierarchy has not only become expressed in the economic relations between landowners and workers, but is also becoming stronger between on the one hand boat owners and boat fishermen, and on the other canoe fishermen. Still, the ideology of inequality and the ensuing practices of domination and appropriation by force have not yet become as hegemonic at sea as on land, because canoe fishermen resist being drawn into unequal economic relationships with boat owners. Canoe fishermen defend their fishing territories against the encroachments of a more aggressive mode of production, while boat owners and boat fishermen press on with their expansion in the name of technical development and social improvement.

Centuries of sociopolitical and economic domination by a powerful class of landowners have always impelled the poor rural masses of Northeast Brazil to escape their dependency. Given the rigid social hierarchy of agriculture, fishing has traditionally been a viable alternative to unrewarding plantation work for those living in the coastal regions. Marine resources are not inherited, one cannot buy titles to fishing grounds, and fishermen cannot be denied access to the sea with laws or other instruments of power used by the ruling class to maintain their privileged position. Fishing seems to place fewer structural and institutional obstacles in the path of poor but ambitious and enterprising men. The common property of the sea is the greatest attraction of fishing (cf. Hardin 1968). However, the belief in an open competition at sea conceals the same deep force relations that pervade the economic organization ashore. Capital owners and fishermen try to monopolize marine resources with all means possible. Their open denial of the ownership of the sea has benefited unregulated forms of exclusion and appropriation. As Marcuse (1964:7) has remarked, "Under the rule of a repressive whole, liberty can be made into a powerful instrument of domination." Technical innovation and capital accumulation have enhanced the strength of boat owners at ousting other fishing modes from their traditional fishing grounds. Sabotage and coercion are legitimized as the progressive thrust of technology and the moral right of higher productivity. This development of sea tenure does not imply private ownership but, not unlike the historical enclosure movements on land, might result in the demise of fishing modes with a lesser concentration of capital and power (Eckert 1979).

This chapter describes the progressive appropriation of natural resources since conquest times and the ensuing social stratification. The history of Camurim can be characterized by an ever-spreading control over natural resources that began several centuries ago with a narrow coastal strip, expanded inland after the extermination of tribes unwilling to yield their forests to the ambitious colonists, and has now reached the shores of the Atlantic. After a brief history of the conquest of Camurim's hinterland, I will continue with an analysis of the booms and busts of cocoa plantations in Bahia and the reign of the planters who instilled an ideology of power, affluence, and autonomy. The development of the regional communication system led to the commercialization of virgin forests and helped to create the salaried labor force of lower class Camurimenses. Finally, I will discuss the principal social groups in the fishing industry in their respective ecological niches, and describe the forms of sea tenure that have resulted from the confrontation of different fishing modes under a growing pressure on the limited marine resources. Paradoxically, sea tenure is a last defense against the complete appropriation of Camurim's natural resources by a small group of boat owners, and at the same time a form of resistance against a mode of production more prone to enhance social inequality and to entail in a loss of individual control over production and distribution.

## *Conquest of the Interior*

Very little is known about the origin of Camurim. Situated only half a day's sail south of Pôrto Seguro, close to where the Portuguese navigator Pedro Alvares Cabral first set foot on Brazilian soil in the year 1500, Camurim must have been explored early on by scouting expeditions and visited by Franciscan friars and Jesuit missonaries.[2] The exact date of the permanent residence of Portuguese colonists is unknown but, most likely, Camurim was a village of Tupiniquin Indians that stood under Jesuit control around the second half of the sixteenth-century.[3]

The Jesuit order was in the forefront of the colonization of Brazil and was as much interested in the material as in the spiritual future of the new continent. Manuel da Nóbrega, who led the first group of Jesuits to Brazil in 1549, writes in a letter of January 6, 1550, to Padre Simão Rodrigues in Lisbon after a visit to Pôrto Seguro and the Indian settlements in the captaincy:

> Until now the merchants and those who came from abroad have not made plantations in fear of being attacked by gentiles. If people would come who would secure the land, then I have faith in Jesus Christ that they will make many [plantations] and will bear as much fruit in the service of God with the gentiles, who will allow themselves to be baptized. It is said that there is here a great amount of gold, and precious stones as well, but which has not been discovered because of the weak force of Christians. (Nóbrega 1955:83; my translation)

The religious zeal with which the Jesuit missionaries tried to convert the native inhabitants to Christianity did not always sit well with the Portuguese settlers. Camurim's location at one of the few rivers in the region, the possible presence of gold, and its richness in brazilwood probably contributed to its foundation. The colonists could employ the pacified Indians concentrated in the coastal settlement as laborers and use them to protect the new colony from the incursions of French merchants (Thomas 1968:52). The Jesuits, however, were too protective of the Tupiniquin to the liking of the secular authorities who were admonished by the missionaries for their moral decline and corrupting influence on the indigenous population. Gaspar Curado, the captain of Pôrto Seguro, even tried to prevent the Jesuits from visiting the Indians. This tug-of-war for the control of the coastal tribes was briefly won by the Jesuits when, in 1591, captain Curado was denounced by the Inquisition for obstructing their evangelical activities. Yet, the tensions kept resurfacing and culminated in the expulsion of the Jesuit order from Brazil in 1759 (Leite 1938:203).

In the sixteenth century, there were many efforts to establish sugar plantations in the region of Camurim similar to those flourishing along the Bay of All Saints at Salvador. Attacks by the Aimoré Indians killed many colonists and their slaves, destroyed the seven or eight sugar mills that had sprung up, and left Pôrto Seguro without inhabitants by 1564 (Soares de Sousa 1940:156–160). The Aimorés offered so much resistance to the invading

Portuguese that most settlements in the captaincy of Pôrto Seguro were destroyed repeatedly in the sixteenth century (Soares de Sousa 1940:159). The third governor of Brazil, Mem de Sá, responded to the continued native resistance with a vigorous pacification campaign in 1558. He attempted to drive the hostile tribes to the interior of Bahia and enlist friendly tribes to fight off the French and their Tamoio allies in Rio de Janeiro, but he could not prevent the continued onslaught on the communities of the captaincy of Pôrto Seguro (Leite 1938:212; Thomas 1968:50–53).

As the Portuguese settlers felt that the cultivation of sugar was too dangerous, they turned their efforts to the search for gold and precious stones. Around 1569, Indians arrived in Pôrto Seguro carrying "news of the existence of green stones in a mountain range many leagues inland" (Magalhães de Gandavo 1922:180). An expedition of 50 to 60 Portuguese and a group of Indians led by captain Martim Carvalho was equipped to search for the mountain of emeralds. The party traveled some 220 leagues inland and found small quantities of gold. Magalhães de Gandavo (1922:182) tells us of their exploits:

> Finally, on account of the enemies whom they feared, and on account of the people who were sick, they turned back again in canoes down a river which is called Cricare; there, in a rapid, was lost one canoe, in which were the grains of gold that they were bringing back as samples. They spent eight months on this trip, and, completely worn out, they returned again to the Captaincy of Pôrto Seguro.

Camurim grew modestly through the sixteenth and seventeenth centuries but never reached the prominence of other early settlements such as Salvador and Recife, which became important ports of export. Demographic maps indicate that the region bounded in the north by the Rio Jequitinhonha and in the south by the Rio Doce was regarded as a no-man's-land, inhabited only by belligerent Aimoré or Botocudo tribes. After the consolidation of the lowlying coastal areas in the seventeenth century, the penetration of Northeast Brazil followed the course of the Rio São Francisco and the Rio das Velhas. Missionaries, squatters, and cattlemen advanced from the north toward the south and established ranches in the Rio São Francisco valley. During the same period, adventurous prospectors (*bandeirantes*) from São Paulo pushed their way north toward the upper Rio São Francisco in search of gold, precious stones, and Indians for the sugar plantations in Northeast Brazil (Poppino 1973:68–112). When they discovered gold around 1693 in the mountains of the Serra do Espinhaço, the northeast became connected with the south through an overland route which further isolated the region of Camurim (see figure 2.1). The Portuguese seemed to have lost their interest in the fabled mountain of emeralds in Bahia as the mines of Minas Gerais proved to be rich in gold and diamonds. The urgency to subdue the Aimoré had faded.

The Aimoré kept attacking and pillaging the settlements of southern Bahia until the mid-nineteenth century, and discouraged the cultivation of lands beyond a narrow coastal strip. Since the beginning of the nineteenth-century, however, increasingly more Indians began to settle near the ranches to work as day laborers and help whites in their efforts to hunt down unruly fellow tribesmen. The Aimoré were among the few tribes to whom

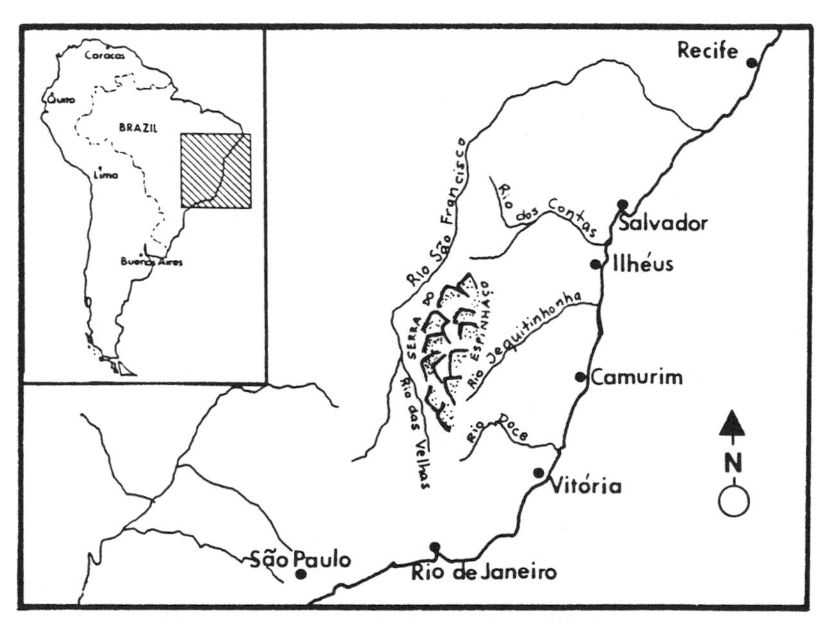


***Figure 2.1***
The Region of Camurim in Colonial Times.

colonial protection laws did not apply as they were greatly feared as cannibals (Thomas 1968:74). "They do not spare any one's life even for an hour, but they very suddenly and swiftly take their revenge; so much so that often while the person is still alive they cut off his flesh and roast and eat it before his very eyes" (Magalhães de Gandavo 1922:110). Great efforts were taken from the sixteenth through the nineteenth century to eradicate them with all means possible, including the deliberate spreading of smallpox.[4] Persistent persecution and internal feuding brought on their eventual defeat.

Native resistance in the rugged hinterland was decisive in the lack of development of southern Bahia, but the shallowness of the regional mangrove-laced estuaries and navigational problems which made it difficult to reach the harbors were also important.[5] Steady southern and northeastern winds obliged ships to tack for long hours to get to their destination, while treacherous coral reefs discouraged many merchants from frequenting the ports. Despite these obstacles, Camurim was considered to be of enough value to receive the status of Town (*Vila*) in 1764 "because it is considered to have enough residents to take advantage of the excellent soil of which it is made, so that these residents can dedicate all exports to manioc and other fruits they cultivate" (*Revista Trimensal* 1896:534; my translation). The Portuguese Crown expressed confidence in the prosperous future of Camurim due to its location halfway between Rio de Janeiro and Salvador, its fertile soil, and its rich, inexhaustable forests.[6]

These great expectations never materialized. Camurim was several hundreds of kilometers away from the major overland thoroughfares that connected Rio de Janeiro and Salvador, and could be reached only by sea. The region supplied timber, manioc flour, and cattle to the metropolitan centers but the total output was far below capacity. The provincial government was concerned about the future of southern Bahia and, around 1850, sent a surveyor to assess its suitability for immigrants. The inspector treated the region in his graphic report as an anemic organ in the body of Brazil. His mission was, as he saw it, to make a diagnosis and provide the government of Bahia with the information "to cure" (*curar*) the illness. He was shocked by the economic and cultural decline (decadência) of a population which still kept manumitted slaves in captivity and failed to make overland contact with the more populous state of Minas Gerais and cultivate

> a remarkably lucrative large strip of land abundant in natural products of easy exploitation, but which is left entirely abandoned to a half dozen dwindling Botocudo tribes who are waging a war of extermination on the audacious and hardworking who take the opportunity to penetrate the area, either from the side of the Province of Minas or from our side, trying to take part of what nature is offering with so much abundance. (*Revisto do Instituto Geographico* 1902:90)

He highly recommends the area for European immigrants and proposes the construction of a railroad and a local network of roads that will link the small coastal towns to the prosperous state of Minas Gerais (*Annaes do Archivo Público* 1923:198–202). "These embryos of life," he continues, "must be interconnected so that, little by little, large arteries will develop whose ramifications form the essence of a prosperous country" (*Revista do Instituto Geographico* 1902:89; my translation).

Notwithstanding the poor state of the regional economy, Camurim was undergoing rapid demographic change. In 1850, Camurim was a settlement of 300 to 350 inhabitants, while in 1899 the population had grown to 1,400 in the town and 5,500 in the county.[7] Three factors contributed to the population growth: former slaves migrated to the coast after 1871 during various legislative stages of the abolition; impoverished peasants fled from the drought-stricken interior from 1877 onwards; and improved maritime transport finally gave Camurim a reliable link to Rio de Janeiro, Ilhéus, and Salvador. The influx of migrants was noticed by the state government, and Camurim was given city rights in 1896. This greater political and administrative autonomy made the local authorities undertake definite steps to lift the county out of isolation, and make it share in the growing economic development of the rest of the state.

In anticipation of its new administrative status, the French engineer Apollinario Frot was contracted to survey the little known hinterland and to prospect for mineral deposits. He was also asked to find the best connection between Camurim and São Miguel (today called Jequitinhonha), a growing farming and mining community in the state of Minas Gerais, 250 kilometers inland. On the 14th of March 1895, the explorer travelled westward from Camurim. In a letter to the authorities of Camurim he complained about the many setbacks of the expedition.

> The majority of our men showed signs of insubordination . . . the forests were extraordinarily difficult to traverse . . . most men fell ill under way, some from fever, and others from wounds to the feet . . . just when we were in a part of the forest where not even birds exist. In the end, we lived eight days on fruits . . . and what fruits!" (*Revista Trimensal* 1895:371–372; my translation)

Apollinario Frot is impressed by the natural beauty of the region and finds the Rio Camurim suitable for shallow water barges and rafts. He proposes the construction of a road to São Miguel that incorporates stretches of the river and navigable lakes (*Revista Trimensal* 1896:199–201). He ends one of his letters with the assertion "Tell Mister Benson that I have found excellent soil for cocoa" (*Revista Trimensal* 1895:373).

The road to São Miguel was never built, but the region attracted the attention of cocoa planters from Ilhéus. The French engineer further mentioned the need for a quay to facilitate the cabotage of goods and suggested the dredging of the river mouth (*Revista Trimensal* 1898:23–24). At the turn of the century, Camurim exported most of its products to Rio de Janeiro and Salvador. Steamships of the Companhia Bahiana de Navegação a Vapor, owned by Lloyds of Brazil, were loaded offshore with coffee, jacaranda, manioc flour, and small amounts of cocoa (*Revista Trimensal* 1898:22–25). Cocoa became the cash crop from which the county of Camurim would eventually derive its greatest revenue and which would form the basis of its current economic development.

## Development of the Plantation Economy

The first cocoa seeds in Bahia were imported from the state of Pará and planted along the margins of the Rio Pardo near

Canavieras in 1746 (Dias 1974:22–36). The soil proved suitable but the hit-and-run Aimoré tribes prevented large-scale cultivation of cocoa. After two centuries of hunting down the independent tribes, the counties of Ilhéus, Pôrto Seguro, Camurim, and other towns were still confined to their residential perimeters. The townspeople were malnourished because not enough food was produced locally and market crops were not grown in large enough quantities to provide for a cash income. Around 1820, however, German immigrants and impoverished Brazilian townspeople infiltrated the dangerous backlands of Ilhéus, Canavieras, and Belmonte to establish small cocoa gardens. These settlers (*desbravadores*) did not rely on slaves to clear the land but were pioneers who conquered the interior with axes, gunpowder, and cocoa seeds.[8]

The period from 1895 till 1930 was the era of the planters' greatest power so vividly described in the novel *Gabriela, Clove, and Cinnamon* by the Bahian author, Jorge Amado. Former desbravadores became influential *coronéis* (colonels) who dominated the countryside, just as the sugar planters in the northeast and the coffee planters in the south had done for many years. But, in contrast, these *coronéis* did not isolate themselves on estates surrounded by slaves but became absentee landowners who lived in town, in order to advance their enterprises through political and commercial means (Garcez and Freitas 1979:19–27). They appointed kinsmen as public officials and exerted political control as patrons of large clienteles (Azevedo 1965:291). The *coronéis* manipulated municipal budgets for better roads, larger harbors, and improved means of transportation to accelerate their expansion, while administrators took care of the day-to-day business on the plantations (Forman 1975:166). The *coronéis* employed seasonal laborers from the overpopulated region of Salvador, subsistence farmers who had fled the droughts that plagued the backlands (*sertão*) in the 1870s and 1880s, and migrants who came from the declining sugar regions of Northeast Brazil. The cocoa planters consolidated their political and economic power during this time and made Brazil into the world's second largest producer of cocoa in 1907, threatening even to surpass coffee as the principal export crop.

The year 1930 marks the decline of the cocoa colonels because of political, economic, and ecological reasons. Getúlio Vargas had seized the presidency after overthrowing the Old Republic and implemented new laws to undermine the *coronéis* and centralize the splintered political forces of Brazil. The professional urban elite became more influential, and the Great Depression in the Western world led to an economic decline hard felt in Brazil (Furtado 1968:213–239).

Destructive cultivation techniques only exacerbated this crisis. Soil exhaustion lowered the production of cocoa because little care was taken to fertilize and preserve the plantations. Cocoa prices on the world market fluctuated greatly, and the financial institutions could not provide enough credit for investments in better methods and new plantings (Filho 1976:89–95; Leal 1976:84–88, 254–258). The establishment of the Bahian Cocoa Institute (Instituto do Cacau da Bahia) in 1931 was the first coordinated effort to deal with the crisis. The institute provided credit, and technical and agronomical assistance to the Bahian plantations, while politically it gave the elite of cocoa planters an institutional organization through which to regain their influence and power in the state and the nation (Garcez and Freitas 1979:79).

The development of the cocoa economy in Camurim ran four decades behind that of Ilhéus. Early experiments with cocoa near the hamlet of Serrania, 50 kilometers west of Camurim, had failed. The settlers concentrated with more success on the cultivation of manioc, rice, sugar cane, bananas, and in particular coffee. The Great Depression and bumper crops in the 1930s made coffee prices tumble and renewed the interest in cocoa, despite the price fluctuations on the world market. The settlers made inroads in the virgin forests and sold tropical hardwoods to the urban centers in the south to generate investment capital. Pedological surveys revealed the same fertile red clay soil (*massapé*) as in the northern cocoa region, and the same hot and humid climate under which shaded cocoa trees thrive.

The political and economic development of the county of Camurim was analogous to that of Ilhéus and its surrounding towns. Poor communications hampered full-scale utilization of the hinterland. Cocoa beans were transported downstream in large canoes that took two to three days to reach Camurim. Sacks of cocoa were stored in warehouses and later picked up by coastal freighters (*Revista Trimensal* 1899:482). These infrastructural circumstances gave rise to cocoa planters who acted as *coronéis* in the rural political structure and were able to attract state and federal funds in exchange for votes (Leal 1976:251).

The final phase in the development of the cocoa economy, both in Ilhéus and Camurim, began in 1957 with the creation of the CEPLAC, the Executive Commission of Cocoa Cultivation Plan (Comissão Executiva do Plano da Lavoura Cacaueira). This organization strengthened earlier statewide efforts of the Bahian Cocoa Institute to teach modern production methods, set up research stations, combat pests and diseases, extend credit, establish schools, and construct railroads and highways (Caldeira 1954:26; Dias 1974:54–60; Filho 1976:94–101). A network of CEPLAC offices was set up in most county capitals, including Camurim, to provide information and monitor the progress made.

The county of Camurim benefited substantially from the programs. New roads opened up previously inaccessible areas, attracting people from other counties and yielding more taxes. However, this very development soon caused the decline of Camurim as a political and economic center of unprecedented growth. The county of Camurim was cut in half in 1960 to create another county around the rapidly expanding town of Serrania. The most powerful and richest men in the region left Camurim and used their contacts and resources to benefit the newly established county. The final blow was the completion of the coastal interstate highway BR-101 between Rio de Janeiro and Salvador in the late 1960s, which made maritime transport of cocoa obsolete and stimulated the rise of boom towns along the new route which led through the cocoa plantations (Bandeira de Melo e Silva 1967:58–61).

Although the cocoa planters now occupy a less prominent place in Camurim than they did a few decades ago, they still furnish a role model for many ambitious Camurimenses. The political and economic aspirations of boat owners can only be understood against the background of a plantation culture in which the landowner lives on the fruits of a secure investment but is far removed from the actual production. The characterization of the sugar planters in colonial Bahia by Stuart Schwartz is still relevant today in the cocoa region: "The preferred goal was to possess landed property, which was valued not simply because it brought prestige in and of itself but also because it was the most secure way to maintain the noble life" (Schwartz 1985:249).

The landowners of Camurim spend most of their time in town socializing with other members of the elite and talking to the prefect, the councilmen, and high-ranking public officials. The prestige and stature they reap from their political and economic influence are most visible in the morning when they meet each other in one of the squares of Camurim. They are dressed informally in Bermuda shorts and lean comfortably on a bench or rest under the shade of a tree. Ordinary people who cross the square greet them graciously but hasten their step, careful not to invade their privacy. This air of relaxed superiority with which they treat the entire town enhances the admiration they receive from the people of Camurim. The landowners give their own slant to what it means to be a man of esteem and power—an example that finds its way into the dealings of men at sea and the ways these men organize their lives.

## *The Social Hierarchy of Camurim*

Camurim receives its identity from the sea, yet the population derives its livelihood mainly from the land. The prefect of Camurim summed up the county's economy in a 1973 pamphlet:

> Her source of wealth consists of the extraction of hardwoods in logs and planks, including jacaranda, as the county of Camurim has the second largest timber reserves of the state of Bahia. She has 1500 registered agricultural businesses which occupy an area of 225,000 hectares with the production of livestock, hogs, bovines, horses, mules, cocoa, cereals, etc.; her cattle herd grew in the year 1973 from 80 to 100 thousand heads and her production of cocoa fluctuated between 10 and 12 thousand sacks of 60 kgs.

Other sources of income in the county of Camurim are fish, manioc, beans, sugar cane, heart of palm, and the production of bricks and tiles. Public services, tourism, and construction provide additional revenues. The elite, which largely controls the municipal economy, consists of cattle ranchers, landowners, owners of medium-size cocoa plantations, owners of brickyards and sawmills, management personnel of financial institutions, upper level public officials, and general practitioners with investments in real estate.

Cattleraising, agriculture, and the extraction of timber offer little employment to the townspeople. Permanent workers live in poor wooden shacks on plantations while migrant workers from the interior of Bahia come to the coastal regions only during harvest time. The manual labor force of Camurim, therefore, has to find employment in the fishing industry, in brickyards and sawmills, or in public service.

Fishing yields potentially the highest income of any manual work available and gives fishermen the freedom to avoid the often authoritarian organization of salaried employment. Fishing is also dangerous, physically demanding, and at times financially unrewarding. Some fishermen earn three or four times the official monthly minimum wage of Cr$23,000 ($46) but others make less than this. A brief description of land-based jobs shows why fishermen reject such employment and prefer to take their chances at sea in the hope of a run of good luck and a better future.[9]

Brick- and tile-makers are the poorest people of Camurim. They get up at dawn and work all day till sundown to earn a measly Cr$10,000 ($20) a month. Brickmaking is unskilled work. A few men dig up clay sediment along the riverbanks of the Rio Camurim and remove the pebbles and stones. They sell the mounds of clay to male tile-makers and mostly female brickmakers. The clay is thrown in a mold, and the bricks and roof tiles are left to dry in the sun for three days. Once thoroughly dry, the bricks are sold for Cr$1 and the tiles for Cr$4 apiece to the owner of the quarry. He fires them and sells them for respectively Cr$5 and Cr$20. Under ideal circumstances, a person could produce 16,000–24,000 bricks or 4,000–6,000 tiles a month, but, in fact, only half that number is produced. Bad weather, lack of firewood, and low demand reduce the earnings significantly. Days of work are lost by an afternoon shower or by prolonged winter rains that fill the clay quarries with water and leave the firewood soaking wet.

Since 1978, the fifteen small brickyards (*terreiros, olarias*) have suffered from the fierce competition of a brickmaking factory that produces cheap hollow bricks (*lajotas*).[10] The owner hires about twenty adults who operate the equipment for a minimum wage. He pays less to the dozen teenagers who cart the bricks to the ovens and drying-racks. Brickmaking is understandably avoided by most adult men with families to support, and only attracts workers who are afraid of the sea, who do not have any contacts to find a place on a boat, or who are itinerant workers temporarily laid off by the plantations. It is almost impossible for men who did not grow up around the harbor or the beach to secure a place in a crew to learn how to fish, because no captain will take the physical and financial risk of contracting an unskilled or unrelated apprentice fisherman.

The construction of the coastal highway in the 1960s and ongoing expansion of cocoa plantations opened Bahian rainforests to commercial exploitation, and turned Camurim into a way station for timber. Truckload after truckload of tropical hardwoods arrives at the three sawmills to be cut into large planks. The sawmills provide employment to about one hundred men who work forty-five hours a week for Cr$23,000 ($46) a month. Machine operators earn Cr$30,000 ($60), while the highest wage of Cr$40,000 ($80) is paid to a few skilled whetters. The work conditions are bad. The sawdust irritates eyes and lungs, and saws are often left unprotected. When one young man accidentally cut off four fingers, he was given one week leave of absence, and received Cr$1,000 ($2) as compensation. No attempt was made to rush him to surgery, and the man was instead grateful not to lose his job. He hung out around the sawmill with his heavily bandaged hand, afraid that someone might take his place. Many families live on the property in company-owned one-bedroom houses and receive free scrap wood for heating and cooking. The companies pay modest insurance and retirement benefits, and allow the men to resell some wood on the side. Some poor people operate a few furnaces on sawmill property to make charcoal.

The emergence of tourism in the late 1970s has provided work to bricklayers, carpenters, and electricians who earn two to three times the minimum wage. These occupations are well regarded but require several years of apprenticeship, and hence are not considered as alternative employment by fishermen.[11]

The last form of employment ashore is offered by public utility companies and the municipality. A handful of men work for the water, electricity, and telephone companies, while at least sixty men are employed by City Hall to carry out menial jobs. Wages range from Cr$9,000 to Cr$42,000 ($18–84). The thirty male and female streetcleaners earn a pitiful Cr$9,000, less than half the minimum wage established by the state government. These men and women live on the same impoverished level of subsistence as brickmakers. Their diet consists of rice, beans, manioc flour, bones discarded at the Sunday market, bread given by a baker, and milk delivered free of charge by a local rancher.

Fishermen are very much aware of the poor labor conditions of salaried workers. Many have worked at least for some time ashore and readily relate anecdotes that describe their bad experiences with employers. These stories from the past are more than self-serving justifications for rejecting such employment but are validated by recurring incidents of abuse and deprivation.

> One summer afternoon, I was on my way to the largest sawmill in town. The dust raised by the trucks settled on my face and I entered a bar to have a drink. Three men were squatting in front of the small wooden shack, one of them with his chest in bandages. "Hey, Azarias," I asked, "what has happened to you?" Azarias, a broad man in his mid-thirties with strong Northeastern features shifted his weight to his right foot and replied, "Well, two weeks ago I was loading a truck at the estate of Odilón. Me and my two buddies here spent the entire morning loading two large trunks of $7\frac{1}{2}$ meters. It took us hours to pull the trunks on the truck with a small tackle . . . and we don't earn more for bigger trees. Odilón pays us only Cr$30 ($0.06) for every meter of timber we transport to the sawmill. When we finally got to the sawmill and I released the cable, the largest piece rolled down and scathed my chest. Odilón had just arrived and he yelled at me that I didn't know what I

> was doing. I tell you, Senhor Antônio," Azarias said to me, "the poor people from here lead the life of the rich man's dog. Odilón threw me some bandages and paid us Cr$450 ($0.90). We went home with Cr$150 ($0.30) each for a full day's work. Do you think that we can feed our hungry children from such a handout?"

Day laborers are not the only men who have been treated badly. Streetcleaners have had their income suspended, plantation workers have been fired without receiving their last week's pay, bad checks have been written, insurance benefits have remained unpaid, and workers have been deliberately harmed and injured in disputes with their employers.

Although fishermen are not necessarily protected from canoe and boat owners who are intent on deceit and exploitation, they at least have more freedom to switch employers. As will be shown in detail in chapter 5, fishermen cherish the intrinsic risk of fishing that is so disliked by landbased workers. Work at sea and work ashore imply different outlooks on the economy. The permanent threat of insufficient means of existence that looms over the heads of all members of the lower class causes an omnipresent preoccupation with subsistence that guides many decisions in life. Yet, despite the need to fulfill similar basic needs of nutrition and shelter, the economic responses and motivations vary significantly. Salaried employees are content with the security of fixed earnings and steady, routine work. They know when they get off work and what to expect in the coming week. They do not like their poor standard of living or the abuses of power to which they are subjected, but they have come to accept such unjust outflows of the class hierarchy as inevitable. Centuries of patriarchal domination created a plantation or "*fazenda* ideology . . . penetrating deeply into all social strata, which seeks to explain the wealth of the wealthy and the poverty of the poor as natural and necessary expressions of intrinsic merits merely sanctioned by a wise and just order of things" (Ribeiro 1971:206). The ownership of capital goods and natural resources is taken for granted as a charter of domination. In the words of one landowner, "I am not poor, I am independent." Poverty and dependence are the inevitabilities of a plantation culture. The only way out of a lower class status is through the help of a patron. Social mobility for the salaried workers is not a quality of Brazilian society but an asset owned by the ruling class.

The fishermen of Camurim are not blind to the differences in wealth in the community but they perceive through the rungs of stratified society opportunities to move higher up. Hard work will eventually reward enterprising men with boat ownership after they have developed the right social connections. Work ashore is therefore not even contemplated by the poorest of fishermen, who may condemn the actions of a too dominant elite but who will not lose faith in the opportunities of social improvement enabled by fishing.

## *The Exploitation of Marine Resources*

The fishermen of Camurim distinguish six ecological zones: the river, the mangroves, the tidal zone, mudbanks, coral reefs, and the edge of the continental shelf (see the appendix for a detailed description). These natural habitats are used by seine fishermen, mangrove fishermen, canoe fishermen, and boat fishermen. Each ecological zone is identified by a unique combination of geomorphic features, sediments, vegetation, depth, and fishing techniques. Given the natural and technical differences that characterize each zone, it is easy for a person from Camurim to guess a fisherman's social and economic position from his fishing location.

Beach seine fishermen have the lowest status in the community, even lower than streetcleaners and day laborers. They are believed to be alcoholics beyond cure, who will go as far as theft to get a drink. The majority of the about twenty middle-aged men who operate the two worn-out beach seines were canoe and boat fishermen in the past. At some point in time, they started drinking excessively and were unable to function as regular crew members.[12] They did not always show up in the harbor at departure time, their production fell, they got into debt, and were eventually dismissed. Beach seining was the only economic activity left for them. An average workday takes only two to four hours; the men can take turns rowing; and the catch is large enough to buy some food and a few glasses of rum. Several men have been abandoned by their families, while others rely on their wives to sustain the household.

The two groups of ten seine fishermen take turns hauling the nets ashore and rowing the six-man canoes. The shore crews always stay on the beach. Shore crews are usually composed of poor women and unemployed day laborers who try to survive during hard times. These crews change every so often because the rewards are small and the work unreliable. Beach seines are not operated in winter and may lay idle for days in summer, due to bad weather or repairs.

Most seine fishermen occasionally catch crabs (*caranguejo, guaiamú, sirí*) in the mangroves in winter and during the tourist season. The crabs are strung into bundles and peddled in the street. Collecting crustaceans can provide a reasonable income of Cr$30,000–40,000 ($60–80) a month, but there are only three professional mangrove fishermen. The mangrove is a hostile environment with snakes, leeches, and swarms of mosquitoes. The dense vegetation is almost impassable and the mud makes walking very exhausting. Most able fishermen are simply not willing to undergo such hardships, while seine fishermen are physically too weak to collect crabs regularly.

Mangrove fishermen also like to fish in the river when large, silvery snooks pregnant with eggs swim upstream to spawn. Unlike seine fishermen, they own canoes that allow them to exploit both the mangroves and the river. They are less successful, however, than the five canoe owners who live near the river at the southern edge of Camurim and use gill nets in addition to handlines. Profits from snook sales are high during the two last months of winter, but pollution and overfishing have reduced fresh water fishing to a seasonal activity. The five southern canoe owners belong to a group of fifty canoe owners and canoe fishermen who derive their livelihood from the tidal zone, and the mudbanks and coral reefs of the shallow coastal waters. These men cast gill nets in the tidal zone in July and August to catch snooks on their spawning run to the estuaries. In summer (September–February), they place the nets further offshore to catch fish that come to spawn at the mudbanks. In winter (March–August) they fish with handlines at the rocks and coral reefs close to shore (see figure 2.2).

Camurim is located near the Abrolhos, a large concentration of coral reefs in the Atlantic Ocean. The reefs have formed on an unusually large protrusion of the continental shelf that at its extreme reaches 240 kilometers beyond the shore—against the average 40 kilometers in neighboring regions. The Abrolhos con-

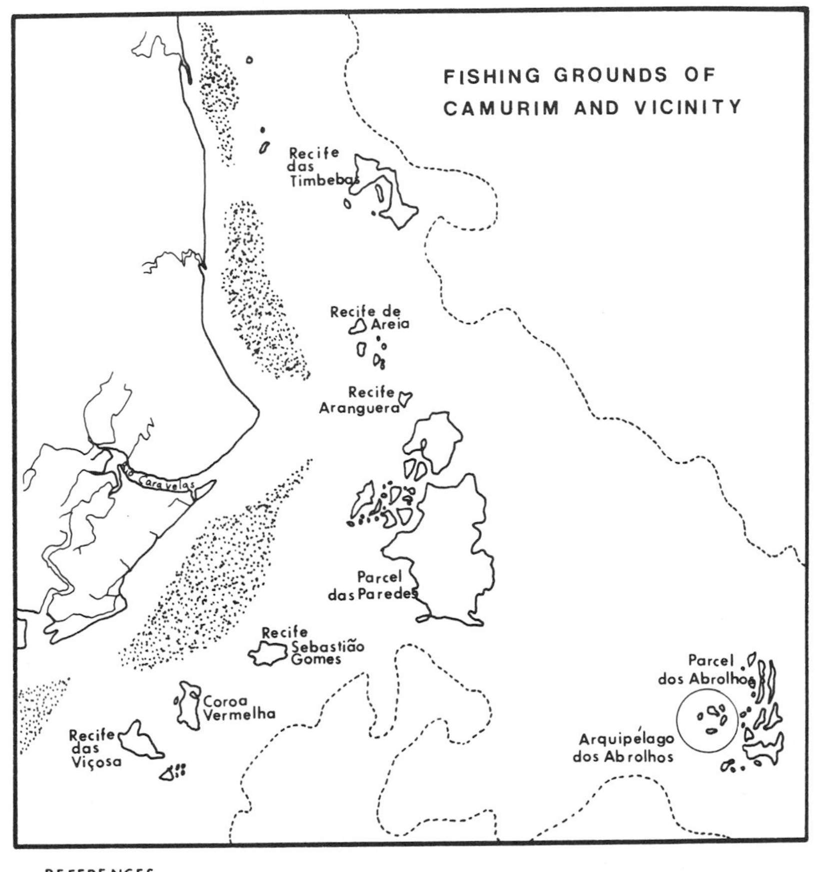


REFERENCES

----- 20 METERS BELOW SEA LEVEL

SAND BANKS

***Figure 2.2***
The Fishing Grounds of Camurim and Vicinity.

sist of two collections of patch reefs with a group of islands in between. The first formation has five large and some smaller oval-shaped reefs and lies at a distance of 10–15 kilometers along a 100-kilometer stretch of the curved shore line. The second formation lies 60 kilometers off the coast. Finally, there is the Abrolhos archipelago of five islands situated 3 kilometers west of the second formation (Barreira e Castro 1981). The main island of Santa Bárbara is inhabited by Navy personnel and has a lighthouse that guides small cargo vessels through a natural channel west of the archipelago and east of the dangerous first arch of reefs. Coral reefs are barely exposed at low tide, while surrounding areas have a depth of 5 to 20 meters.

The twenty-five boat owners and one hundred and twenty boat fishermen in Camurim work with liners, which fish exclusively with handlines, and multipurpose boats in different ecological zones. Large motorboats make the eight to fifteen hour trip to the second coral reef formation in summer, and in winter fish at parts of the continental slope which are about 80 kilometers from Camurim. Small and medium-size boats use handlines in winter from May till August around the first arch of coral reefs (as far as five hours away), trawl for shrimp between November and February at the same mudbanks frequented by canoe fishermen, and go netfishing during the rest of the year.

On a social scale, the lowest rung is occupied by the seine fishermen, the mangrove fishermen reach slightly above, and the boat owners are at the top. Canoe owners, canoe fishermen, and boat fishermen are in disagreement about their relative position on the middle echelons. Canoe owners and canoe fishermen are convinced that they are better off and have more prestige in the community than boat fishermen, while boat fishermen believe the opposite. Most canoe owners even argue that they are superior to boat owners who are often in debt to financial institutions and fish dealers.

These disputes about social rank are at the heart of an economic activity organized around fluctuating revenues reaped from a natural environment that precludes any legal claims to ownership. If canoe fishermen, boat fishermen, and owners of boats and canoes want to maintain their self-perceived high rank, then they must secure a firm resource base and contest yet unexplored ecological zones. Declarations of superiority serve indirectly to legitimize such encroachments. Accusations of overfishing versus underutilization, of destructive technology versus outdated fishing methods, and of ritual neglect versus superstition are presented as justifications for either the preservation or the elimination of traditional ecological niches. The protection of natural resources becomes essential to maintaining one's socioeconomic status, and every group and individual has developed ways which reduce competition and enforce the boundaries of its fishing territories.

## *The Enclosure of the Sea*

The captains and fishermen of Camurim do not know how to read nautical maps and they do not have the navigational implements to utilize them. Instead, they use a widespread system of visual triangulation that arrives at the coordinates of fishing spots by lining up several landmarks (cf. Forman 1970:65–74). There are three types of landmarks: *montes* (mountains), *moitas* (high trees), and *marcas* (large constructions such as lighthouses, churches, and watertowers). Marcas and moitas are used close to shore because the mountains that are 30 to 300 kilometers inland cannot yet be seen. Montes and moitas identify fishing spots further offshore. This system of alignment is, of course, only used by fishermen who fish within sight of land.

Liners that operate in the open sea have developed a navigational system which is not as accurate as the landmark triangulation but that still allows them to reach specific areas at sea. They use a compass and a wristwatch as spatial and temporal markings. Coral reefs are not suitable as landmarks but they give fishermen some sense of direction. For example, this is how the captain of a liner traces the route to the Abrolhos archipelago: "Go for three and a half hours in a south-southeast direction until you arrive at the northern tip of the Parcel [a large coral reef], proceed one hour to the south, and finally turn east until the lighthouse of the island of Santa Bárbara becomes visible."

When fishermen arrive close to a desired spot, they drop a

plumb line to measure the depth and to feel by the shock reverberating through the line if the sea floor is rock, sand, gravel, soft, or firm clay. Sometimes, they attach a piece of soap to the plumb to examine the sediment and identify the spot through smell and color. The importance of these elaborate ways of ascertaining the exact location of fishing grounds may be exaggerated, as Kottak (1983:96) notes, to protect the prestige of certain captains, but they are crucial to increasing catches in a complex, diverse, and—above all—invisible natural environment.[13]

The lack of precision in locating fishing spots, the vastness of the sea, and the relative randomness with which fish are caught make the exclusive ownership of marine resources impossible. However, in response to declining catches (cf. McCay 1978), captains attempt to limit the access to marine resources through secrecy about navigational methods and fishing grounds. Secrecy is prevalent among boat fishermen as well as among canoe fishermen. Both use the same tactics to mislead other fishermen. A captain will pass over his destination and drop anchor elsewhere when he suspects that another boat is following him. He will only return when the competitor is out of sight. If a vessel happens to surprise another crew in the middle of a good catch, then the fishermen will not retrieve any hooked fish until the boat leaves. Some canoe fishermen confessed to suspending the catch from a line in the water to deceive their colleagues.

Open conflicts about sea property rights, instead of mere secrecy, exist between canoes and trawlers in shallow waters. Unlike netfishing boats that are usually in waters beyond the reach of canoes, trawlers are shrimping in clay banks 100 to 1,000 meters offshore. In summer, ten to fifteen boats drag trawl nets back and forth over the sea bottom on a stretch of about 7 kilometers in front of Camurim. When boats began to trawl for the first time, there were frequent incidents of boats ruining gill nets on their course. In order to make the nets more visible, canoe fishermen attached bamboo poles with red and black flags to the middle and the ends. The poles give the impression of stakes protecting private property against intruders, of legitimizing ownership in a frontier territory. The damage of nets diminished but did not cease altogether.

> One day in December 1982, I was observing several trawlers tacking within 100 meters of the beach when I saw one boat cutting through the gill nets of Zé Silva, a canoe owner. The shock of the impact was so great that one of the boat fishermen fell overboard. In the afternoon I asked the two men on the crew what had happened. "Well," replied one, "we must have dozed off while we were resting on the stern and we simply didn't see the flags." I asked them if they were going to pay any damages to Zé Silva. "Certainly not, he shouldn't be there with his nets." A few days before, I had been told by an angry trawler owner that he had deliberately cut through some gill nets. "Those fishermen of the north cast their nets right in the middle of our trawl routes," he said, "and I just passed on top of them. The sea belongs to everybody. I am stronger, so they should put their nets somewhere else."

The tension continues and canoe fishermen try to protect their nets by placing them behind submerged rocks that are avoided by trawlers. Damaged nets are hardly ever reimbursed. Boat owners simply deny that their boats did the harm and the absence of everyday interaction between the two groups make it difficult to make claims. "We do not receive any compensation for our destroyed nets because the lieutenant of the Navy is on their [the boat owners'] side," complained a canoe owner. "When he comes to Camurim to look into these matters, the President of the Fishermen's Guild buys him off."

There are also occasional conflicts between—on one side—trawlers and canoes, and—on the other—beach seines. The combined exploitation of the coastal waters by these three fishing methods puts too much strain on the habitat and has led to overfishing. Shrimp trawlers remove a crucial link in the marine food chain, tiny mesh trawl nets kill many immature fish—eventually reducing the number of reproductive adults—and gill nets prevent migrating species from approaching the coast to spawn, thus reducing catches in the tidal zone. Furthermore, canoe fishermen place gill nets wherever they want to with no regard at all about the casting pattern of the beach seines, while trawlers with their noisy engines disturb the fish close to shore. However, the legitimate complaints of the seine fishermen are simply ignored because they do not have enough physical power and social status to affect the fishing strategies of others.

The relationship among shellfish collectors is without conflict because the mangroves are still rich in crabs and mangrove fishermen can visit areas by canoe that seine fishermen cannot reach on foot. Intense disputes about fishing territories among canoe fishermen will be discussed in the next chapter.

The overall decline in catches by boats and canoes is causing an intensification of the production effort with more and better implements, a diversification of fishing methods, and a further expansion into each other's waters. Owners of small boats are enlarging their vessels to a medium size of 8 to 9 meters that seems best for trawling, netfishing, and handlining.[14] However, these strategies only further endanger fish populations and will eventually lead to diminishing returns. A competition for scarce resources has begun that signifies more than a runoff between different fishing strategies. It is a conflict between different interpretations of nature and society, interpretations that are intimately related to the economic practices.

Before the presence of boats in Camurim, the fishermen believed that the sea was without end and its resources unlimited. "We used to catch so much," reminisced Elmício, "that my mother would beg me not to bring any more fish because she didn't know what to do with it. My nets were so full that they would burst when I tried to haul them in all at once." The arrival of boats confirmed this abundance as their catches greatly surpassed previous records. This belief was not shaken when catches declined after some time because larger boats, more powerful engines, better nets, and longer fishing trips to unexplored grounds made the productivity improve immediately. However, as catches failed to remain high and the crews from Camurim felt the competition at sea from numerous large trawlers, Japanese longliners, and handlining vessels based in Vitória, Macaé, Cabo Frio, and Rio de Janeiro, which together brought in hundreds of tons of fish with the aid of detailed nautical maps and sophisticated sonar scanners, they began to realize that fish seemed to be scarce after all.[15] Overfishing completely turned around their conception of nature by revealing the finitude of its resources.

Unable to affect the fishing practices of the many vessels on the high seas, criticism was directed inwards, and boat fishermen began to place the blame on canoe fishermen. One group of boat fishermen reproached the canoe fishermen for using destructive fishing techniques and for overfishing the coastal waters. They were told to abandon the small mesh nets and stay out of the shallow waters during the spawning season. A second group instead accused the canoe fishermen of underutilizing the rich marine resources by using outdated fishing methods, preventing boats from trawling in certain shallow areas, and by being unable to take advantage of fertile grounds further away. Both groups of boat fishermen acknowledged the scarcity of fish but suggested a more balanced exploitation.

The reaction of the canoe fishermen to these accusations and the declining catches was surprising. Some charged that the sea was as abundant as ever but that the fish were getting smarter. They had learned to detect and avoid the barely visible nylon nets. The result was that these canoe fishermen intensified their fishing effort with more nets to outmaneuver the elusive fish.

> One afternoon, Cassimiro told me that Francisco and Murílio had caught 80 snooks in their nets. I asked him, "Cassimiro, but aren't they saying that the sea is coming to an end?" "No way," he answered, "how many people aren't there in the world? Well then? A woman can have only one child a year. And even so, with as many people as there are, we cannot occupy all the space of the sea with all these people. A fish has every time thousands of eggs when he spawns. I have seen fish full of eggs. It just can't come to an end. There is so much place in the sea to hide. Fish are just like people. They gather in one place. They run in schools. When people leave the house to work, they return to eat, to sleep, and to be together. They come from and go to a

place, just like fish. When you cannot find fish in one place, then they must be in another. So when it seems that they have completely disappeared from one place, then it is because they are somewhere else."

Another group of canoe fishermen argued that the fishermen of Camurim had for quite some time failed to perform the proper offerings to the Sea Goddess. The Sea Goddess was offended by their laxness in carrying out the age-old rituals and was punishing the fishermen with diminished catches.[16] They made some oblations at a local shrine, took presents for Iemanjá to sea, became more sensitive to the ritual observances of the calendar, but did not change their fishing methods. A growing group of canoe fishermen, however, have begun to blame the lower productivity on the use of beach seines, trawlers, and small mesh nets. They are suspicious of formal regulations to curb the overexploitation of the shallow waters, and are using social control and territoriality to protect the scarce marine resources.

The discursive conflicts about overfishing and the ensuing territorial disputes have also revealed a different interpretation of its social consequences. The most ambitious boat fishermen have been arguing that canoe fishing and beach seining have thwarted the growth of the motorized fishing fleet and, as a result, their socioeconomic mobility. The demise of alternative fishing modes will raise their income and will enhance their chances of owning a boat.

The clash of fishing modes has a different meaning for canoe fishermen. They claim to see through the false front of self-serving promises and warn young men with less strong convictions in canoe fishing that "the boat offers a baited hook. Whoever swallows it will not only lose his independence as a fisherman but will also lose control over his family." Boat fishing is an expansionist mode of production which monopolizes marine resources and makes crews dependent on the boat owners. The sea becomes the property of the men who own the capital means to harvest it and who, therefore, determine the conditions of production and reproduction. The eagerness of boat owners to pursuade skilled canoe fishermen to join their crews and the, at times, ruthless ways with which they beleaguer the traditional fishing grounds of canoe fishing reinforce this interpretation.

Cordell (1989) has argued that the control of marine resources—defined by Brazilian law as common property—is a source of pride to lower-class men who live in a society where land ownership is beyond their reach. Yet, behind these poor man's riches appears the same ideology of inequality and domination that has for centuries guided the cultivation of land in Northeast Brazil and threatens to steer the exploitation of the sea in the same direction. Treating the rainforests as virgin and yet unclaimed land, the planters and landowners of Camurim expropriated the native inhabitants in order to grow a monoculture of export crops and obtain an absolute hold over the social, political, and economic destiny of their laborers. Boat owners have emulated these hegemonic practices, and are trying to command and monopolize the resources of the sea to the exclusion of alternative modes of production, behind the guise of the common property of the sea. Ironically, the only way that canoe fishermen can hope to continue their way of life is by practicing the same strategies of control and exclusion which they condemn as immoral and unjust.

# 3
# Canoe Fishing Along the Atlantic Coast

When I first arrived in Camurim I was urged to talk to Pedro Moisés because he could tell me everything I would like to know about canoe fishing. Moisés was indeed a very articulate man whose years of adult education had allowed him to say farewell to the arduous work at sea and accept a position as assistant to the county's land surveyor. His job was to drive the surveyor's jeep, set up the equipment, write down the measurements, and occasionally mediate land disputes among the smallholders in the county. Given the good pay and the ease of his tasks, his former fishing colleagues were surprised when Pedro Moisés resigned after less than a year and returned to canoe fishing. I asked Moisés why. "In fishing," he said as we walked along the beach, "I am my own boss. I can go to work whenever I want to, and stop whenever I want to. I am independent and owe no responsibility to anyone." He complained about his job as an assistant surveyor. "It's true that I had the respect of the fishermen, the people of Camurim, and the smallholders I talked to, but I also had to swallow the orders of the surveyor, the accusations of being corrupt, and the pressure from the plantation owners to decide disputes in their favor." "But why return to fishing?" I insisted. "Now, I don't have any *patrão* (boss, master)," he replied. "Now, I am the *patrão* of the fishing spots which only I know."

The canoe fishermen of Camurim see themselves as free, independent, and autonomous men who live in a highly stratified society which continuously threatens to absorb them into its web of hierarchical relations. They believe that canoe fishing gives them a greater independence from the oppressive relations of the plantation economy than any other employment in Camurim. Yet, there is considerable controversy about which are the proper fishing practices to protect that independence. The majority of canoe fishermen regard their organization into corporate groups to defend fishing grounds assailed by overambitious colleagues and expansionist boat owners as the most effective form of cooperation to secure their status as independent producers. A minority of nonaligned canoe owners sees it as sacrificing the autonomy that is the very essence of their existence.[1] The latter accuses the former of betraying their identity as canoe fishermen, while the majority accuses the minority of refusing to defend their way of life. Aside from providing an ethnographic description of canoe fishing in Camurim, this chapter shows how these diverging interpretations about social identity lead to tensions between the two groups of canoe fishermen and result in different economic practices.

Practice theorists such as Bourdieu (1977:92–95) and Giddens (1979:120–130) emphasize that identity is not an immutable property of individuals but is continuously reproduced in practice. The formation and transformation of identity are lifelong social processes turned towards the self.[2] Identity originates in childhood practices and is acquired through socialization processes that have been analyzed by George Herbert Mead (1977:199–246).[3] Identity is as much a result of intersubjective mirroring as of self-definition (Dundes 1983:238). The identity formation of most canoe fishermen in Camurim began during childhood and adolescence as they went fishing with their friends in the lakes nearby and accompanied their father or elder brother at sea. They learned to maintain their balance in an unstable canoe, to bait hooks, reel in fish, mend torn nets, and were made fun of when a sudden clumsiness betrayed their inexperience. They also began to interpret their social world through the perspective of the adult fisheremen, and began to understand their own actions in a similar way. Finally, they accepted the discourse of the group towards canoe fishing as their own and identified themselves with the professional fishermen (cf. Blumer 1969:12–20; Cicourel 1970:147–153). They recognized their growing mastery as an indication of their successful emulation of the social identity of canoe fishermen, and felt with them a sense of sameness and continuity (Erikson 1959:22–23). They realized that they shared certain basic characteristics with the other canoe fishermen, that they had acquired their social identity.

The social identity of the canoe fishermen of Camurim has also become profoundly influenced by their comparison of boat fishing and canoe fishing. Canoes have become cultural foci of freedom, independence, and endurance for owners and fishing partners alike. This overlap between the attributes of social identity and the interpretation of the canoe as a cultural focus is not accidental. The social identity of the canoe fishermen is a mirror of the social circumstances in which they see themselves. What type of persons they become depends to an important extent on the conditions under which they can maintain canoes as their principal means of production.

The expanding social significance of corporate groups that regulate the exploitation of coastal marine resources demonstrates how identity and discourse are intertwined in economic practice. Sea tenure has branched out into forms of social interaction that in the past were considered to be clearly beyond the economic realm. Fish is redistributed to fishermen who have had a poor catch, assistance is given to ill colleagues, and leisure activities are commonly shared. Due to a rift between nonaligned canoe fishermen and fishermen organized in corporate groups, considerable disagreement has arisen about whether or not the canoe fishermen of Camurim have lost crucial attributes of their social identity which in the past always distinguished them from members of other professions in town. Nonaligned fishermen have been arguing that the corporate members still perceive themselves as autonomous and independent, but that they no longer have the social identity of their fathers and grandfathers, who braved the seas singlehanded and went wherever their intuition drove them. They believe that sea tenure has affected the way the corporate members behave towards one another and has changed their social identity as canoe fishermen. The associated fishermen counter that their nonaligned colleagues have lost their social identity because they fail to fight for the continuation of their life style and have resigned themselves to the fatal encroachments of the boat owners. Nobody disputes that sea tenure has certain material advantages or that the essence of a canoe fisherman is to be autonomous and independent, yet the two groups disagree on who still retains that social identity and what practices can guarantee it.

I will begin with a description of sea tenure in Camurim and analyze the open conflicts between corporate groups and nonaligned canoe owners. After a discussion of fishing techniques, the division of labor, the catch division, and the marketing of fish, I will proceed with an analysis of netmending parties. The mending of nets is strictly a productive activity but the conversation between menders and onlookers shows how fishermen interpret economic practice. Joking relationships with aspirant fishermen define the conditions of the corporate groups and socialize teenagers into the privilege of membership. The canoe fishermen emphasize the importance of work in reproducing society and expose teenagers as deficient in their social contribution. I will conclude with a discussion of the conflicting interpretations about the social identity of canoe fishermen and the importance of the canoe in safeguarding their independence and social equality as producers.

## *Sea Tenure and Corporate Groups*

Sea tenure did not exist three or four decades ago when a small group of canoe fishermen could often not even sell the fish caught in one or two cotton gill nets. Camurim was much smaller, the fish market was restricted to the local population, and the cabotage of cocoa was still one of the main sources of income. However, the demand for fish rose rapidly in the mid-1960s as flourishing cocoa cultivation caused an economic boom which attracted thousands of plantation workers and their families from the semi-arid interior of Bahia. As cocoa was sent by truck to the export harbor of Ilhéus, most Camurimenses who used to work in the river transport switched to fishing. With a growing outlet for fish and a larger group of fishermen ready to satisfy the demand, wealthy townspeople invested in fishing gear and began a virtual onslaught on the rich coastal waters with beach seines, gill nets, and eventually with boats. The productivity of gill nets rose as cotton thread was replaced by stronger and less visible monofilament nylon line. This important innovation and the simultaneous proliferation of fishing gear led to overfishing. During the same period, population grew rapidly—from 3,685 inhabitants in 1970 to 5,708 in 1980—adding many young men to the fishing economy. Population pressure, better nets, an expanding motorized fleet, canoes with a limited operative range, declining productivity, and a growing market made canoe fishermen more protective about the dwindling shallow water resources. In an effort to maximize returns, some began to move along the coast in search of better but more distant fishing grounds. They moored their canoes at these new locations and began to feel that they had a right to ban newcomers, and deny them access to marine resources which were in principle the common property of all fishermen.[4]

Walking the 8-kilometer stretch from the mouth of Rio Camurim in the south to the lighthouse in the north, one finds the first sign of sea territoriality in the clustering of canoes at different places along the beach. These clusters represent five groups of canoe fishermen who control parts of the beach and the adjoining fishing spaces. The groups are known as *turmas* and are identified by prominent landmarks that provide a name behind which the members can range themselves. The *turma do clube* lives closest to the city near a popular dance hall (*o clube*) that faces the beach of Camurim. The *turma da Lagoa Doce* remains near a lake 500 meters north of the *clube*, the *turma da Lagoa Comprida* stays about 3.5 kilometers further to the north at a second lake, and the *turma do farol* beaches its canoes at the base of a lighthouse 6 kilometers north of the city, at a one-hour walking distance. The only group in the south is the *turma do rio* which fishes in the river and around the estuary. The turma do clube is the largest group with fourteen members. The other four groups consist respectively of eight, ten, nine, and five canoe fishermen. The first four groups reside in the northern part of town. Members of the turma do rio live near the harbor at the southern edge of Camurim.

Fishing territories are ideally limited by easily identifiable landmarks such as cliffs, sand dunes, palm trees, and houses, but turmas disagree about the proper demarcation of the fishing areas. Their boundaries are not permanent but shift through recurring verbal and physical confrontations betweeen neighboring turmas.

> Casimiro sensed some serious trouble when he saw Cipriano beach his canoe empty-handed. Their gill nets had been cast awfully close during the last two days but neither one seemed willing to budge. Cipriano lifted the oar on his shoulder, turned around abruptly, and assumed a threatening pose in front of Casimiro. "Why are you casting your nets so close to me? Are you trying to catch the snooks away from my nets? Don't you know that my turma has established rights in this area?" Casimiro took a step forward and looked at him defiantly. "You're a cuckold," he said, "and I'm going to call my friends to decide who controls this area." Cipriano pulled his knife and hurled some insults at Casimiro, but he quickly backed off. When Casimiro returned the next morning with a fishing partner, Cipriano had already removed his nets.

Canoe fishermen never allow these confrontations to lead to a standoff because, more than physical violence, they fear that the other group will take fish from the trespassers or even steal the valuable nets themselves. The theft of nets seldom occurs among the fishermen of the north but happens almost every year between the turma do clube and the turma do rio which share a disputed boundary that is situated somewhere between the town and the estuary in the south. The two groups live in different neighborhoods and hardly ever associate with one another. Stolen nets would be difficult to trace after the floaters and main cords have been changed.

The manner in which disputes between corporate groups are resolved is an indication of their social distance. Northern canoe fishermen have occasional verbal quarrels but they avoid more extreme acts of aggression among themselves because of their greater social intimacy and the likelihood of detection. They live in the same neighborhood, they see each other daily on the beach, exchange formal greetings, and might share some distant kinship ties. Furthermore, territorial claims are less vigorously defended in winter when net fishing is uncommon and the canoes fish farther offshore with handlines. During this period there is no advantage in maintaining the separate clusters of canoes along the beach but, instead, the northern fishermen often place their canoes close together near the town or the lighthouse to protect them against bad weather and spring tides.[5]

Sea tenure is most forcefully upheld in summer and at the end of winter when canoe fishermen use gill nets. Summer is the spawning season and fishermen cast nets in those locations where they assume the fish have advanced on their way to the coast. Territoriality is especially strong in January and February when the fish are closest to shore, and during the months of July and August when snooks or *robalos* (*Centropomus ensiferus*) ascend the river so spawn upstream. Gill nets are placed perpendicular to the shoreline and near the river mouth. The nets are clearly identifiable from the beach by their colorful flags, so that each turma can monitor the movements of adjacent groups, and ascertain if some are violating their fishing space.

To constitute corporate groups, turmas have to have more in common than the control of some vaguely delimited fishing grounds. Anthropology has defined a corporate group as a group of people who operate as a legal individual under one name, manage an estate that is cooperatively exploited and defended, and whose positions in the social structure have a degree of continuity by replacing old with new members (Keesing 1975:17). A turma does not have a legal status that gives it property rights to the sea but the members act together to secure the estate through forms of litigation that are practiced by other turmas.

The members of corporate groups in tribal societies often trace their descent from a common ancestor to guarantee their access to the estate. Turmas are not descent groups, they do not have exogamy rules, and are therefore more like open corporate groups that allow individuals to join or leave after the consent from influential members (Radcliffe-Brown and Forde 1965:42). Turmas persist, therefore, through the voluntary retirement of old fishermen and the recruitment of young adults.

Still, despite the absence of linear descent, each turma has a core of consanguineal and affinal relatives who together form a kindred-based group. The core members of the turma do clube are four brothers and two brothers-in-law, the turma da Lagoa Comprida consists of a father, his two sons, and the father's brother, while the other three groups have similar arrangements of relatives. Aside from these close kinship ties within each core group, the members of the four northern turmas often share some distant kinship ties through their grandparents or great-grandparents, but this common descent is not called upon to claim access to other corporate groups and their fishing territories.[6] The only turma that does not have any traceable kinship relationship with the members of other corporate groups is the turma do rio. These canoe fishermen live near the harbor in southern Camurim and are related to the boat fishermen in the neighborhood.

Residence is a more important unifying characteristic of corporate groups than kinship. The members of each turma live close to one another, often in the same street or block.[7] The daily exchange of information about the possible whereabouts of fish and the movements of other corporate groups is more crucial for a territorial claim to fishing grounds than affinity and consanguinity. Fishermen who live in the same part of a neighborhood have more common interests and can share valuable information more easily than relatives who live apart. Close kinship ties with the core members of a corporate group may facilitate recruitment, but residence is a more important ground for membership.

The members of a corporate group do not unite into one production unit to catch fish, nor do they entirely share their personal knowledge of rich fishing spots. Crews may help one another in pointing out areas where migrating species are hiding, but in general they do not reveal their fishing strategies or the exact location of their destination. This limited cooperation extends itself also to the use and maintenance of means of production. Corporate members help to mend each other's nets when these are heavily damaged, and they will clean and protect each other's canoes when the owner is ill or traveling. This mutual care for the means of production enhances the social cohesion of the corporate group and springs from a collective responsibility for the material wellbeing of its members. Canoes are therefore loaned to fellow turma members whose dugouts are undergoing repairs. This obligation, however, does not bear upon the use of gill nets. Gill nets are expensive, easily damaged, and a person simply will not reduce his income by loaning them to others. Fishermen regard corporate groups as a beneficial way to secure resources but they are not willing to make sacrifices that are in conflict with their personal interests.

The solidarity of turmas is also expressed through social relationships that are not primarily production related. Turma members gather around colleagues who are mending their nets, or hang about on street corners to discuss the events of the day. At night, they demonstrate their corporate allegiance by watching television together, or playing cards and dominoes. Such recreational activities strengthen group cohesion and add to the multiplexity of role relationships among the members.

Another expression of corporateness is the redistribution of fish among members. Corporate members do not pool their catches, but a crew that has caught nothing can always rely on help from more successful colleagues. The turma gives a form of social security to its associates. The economic value of the collectively dispensed resources is very small in relation to the total amount of fish caught annually, probably less than 5 percent, but the allocation is a highly significant demonstration of group solidarity. The redistribution does not eliminate income differences among canoe fishermen, but it strengthens ties within the group.

Sea tenure, close social ties among turma members, and the existence of an allocation system have changed the fishing practices of these canoe fishermen. They have begun to specialize in the use of wide-mesh gill nets (*caçoeiras*) which reduce the cost and labor needed for their maintenance, maximize profits, and diminish the risk of overfishing. This specialization has also enhanced the mutual dependency of the members, and assured the continuity of corporate groups by using gear that conserves coastal fishing stocks. Wide-mesh nets are stronger, need fewer repairs, and catch larger, more profitable fish than fine-mesh nets (*tainheiras*). The drawback that the nets may fail to catch anything is offset by the redistributive exchange of fish. A canoe owner who wants to cut all ties with his corporate group will therefore be obliged to buy fine-mesh nets that always guarantee some fish.

There are four canoe owners who do not belong to any of the five turmas. These men live among the northern canoe fishermen but avoid all contact with one another and with the turmas. They are very independent and withdrawn men who, as one said, "dislike joking and gossiping on street corners, and do not want to be taken advantage of by those lazy colleagues." Their independence forces them to catch whatever they can with fine-mesh nets, for they are not part of the redistribution networks. To a certain respect, it would be detrimental for these men to be part of a turma because they would be more often called upon for help, while they themselves would never have the need for reciprocation.

The turmas complain about the "predatory attitude" of the nonaligned fishermen, accuse them of depleting the marine resources, and blame them for the overall decline in productivity. The men are tolerated as long as they do not hinder the movements of the turmas. There have been a few conflicts and net thefts but, as there are only four of them, their aggressive stance in defending the common property of the sea can be tolerated. The four canoe owners pay for their independence with higher maintenance costs, more work, fewer social benefits, and lower productivity. They do not generate enough profit to expand the array of fishing gear with wide-mesh nets, and their choice of fishing nets forces their families to a lower standard of living than the members of corporate groups. They willingly pay this price to maintain the independence that attracted them to canoe fishing to begin with. "I prefer to work eight hours a day, ruining my eyes and my back in front of these nets," said Natalício, "than to give up my freedom by associating myself with those turmas."

The only other persons who can go across territorial boundaries with impunity are two half-brothers who are affiliated with the turma do clube. They owe this privilege to their great generosity. They are the two most successful canoe fishermen in Camurim, who liberally share their good fortune with anyone in need, and who have established redistributive exchange relationships with other turmas. Their benevolence is rewarded by those corporate groups with permission to enter their fishing grounds. They have won through cooperation what the four nonaligned men achieved through antagonism.

## *The Technology of Canoe Fishing*

Canoes are made from trees extracted from the coastal rainforest of Camurim.[8] Although there is not yet a shortage of large trees, the unrestrained exploitation of the forest is rapidly raising lumber prices in Camurim. Whenever possible, fishermen avoid dealing with sawmills and try to negotiate lower prices with local landowners.

Fishermen make a distinction between a *batelão* and a *canoa*. Batelões are a bit larger (0.66–1.10 meters) than canoas and have higher prows, sterns, and boards to make them more seaworthy. Canoas are best suited for rivers and mangroves where the water is more calm. A canoe that is regularly painted, beached carefully, and protected from the sun with dried palm leaves, will last up to twenty years. Most canoes are made to be used by one or two persons, to allow its owner the choice of fishing independently or with a partner.

A canoe builder takes two to three weeks to make a dugout canoe (*batelão*) that is about 0.75 meter wide and 5 meters long, and charges Cr$12,500 ($25) a week for labor. There are only two old canoe builders in town, who also work as shipwrights. Fishermen who do not want to wait till the builders have time available, or who do not want to deal with local landowners and sawmills, buy canoes in the small villages north of Camurim that still fish exclusively with dugouts. Once a canoe is fully equipped with a large lateen sail, a small jib, a detachable leeboard, two benches, two oars, a killick, and ropes, it costs around Cr$75,000 ($150). The price of a used canoe varies according to its quality of wood, shape, age, size, balance, and overall condition. A ten-year old canoe may sell for as little as Cr$15,000 ($30).

Sail-driven canoes have a very limited operative range that obliges fishermen to optimize the exploitation of shallow waters with several fishing techniques.[9] All fishermen use handlines, set lines, and gill nets. Set lines (*linha de espera*) with large baited hooks are left overnight at sea, tied to a killick and a float. Early in the morning, the fisherman takes the fish and replaces the bait. There are three main types of handlines. The most common handline (*linha parada* or *linha de fundo*) consists of a line, a lead weight, and one to six baited hooks that catch fish near the sea bottom. Trolling lines (*linha de corso*) have very light sinkers and are towed through the water on the way to the fishing grounds. Floating lines (*linha de bóia*) do not have leads because these lines are intended to float on the surface of the water in strong currents.

Effective as this cheap gear is during certain periods of the year, a canoe fisherman cannot make a living from coastal fishing without owning gill nets. A gill net is 60 meters long and 2.7 to 5.4 meters high. Together with floats and leads, a new ready-made net costs Cr$42,500 ($85). These nets are for sale in Camurim and in any city nearby. Some fishermen have made a few nets themselves, but the modest savings do not compensate for the substantial investment in labor.[10]

Canoe fishermen use only stationary nets that are kept in place by killicks. Depending on the ratio between floaters and sinkers, a gill net captures either pelagic or demersal species of fish. More sinkers and fewer floats will make the net rest on the bottom of the sea, while a reverse proportion will make the net catch mid-water and surface fish.

A canoe owner must have at least seven gill nets to operate as an independent producer. A single net, or one set of several nets, does not yield enough income. Furthermore, gill nets are never cast together but are placed in groups of three or four at different locations to diminish the risk of loss and to enhance the chances of catching fish. The total capital investment for a fisherman who wants to own a canoe, fishing lines, and seven nets is about Cr$375,000 ($750). A fisherman never buys these capital goods all at once. As a bachelor who still lives with his parents, he first buys a few used nets. The income from his catch is not spent in the household, but is reinvested in new nets. Gradually, he accumulates a half-dozen nets and decides to buy an old canoe. He will not procure the help of wealthy townspeople, but prefers to finance the capital goods with his personal savings.

Celestino woke me up at dawn. I took a light breakfast of tea, boiled plantains, bread and cheese, and went to his house. His eldest daughter was warming yesterday's coffee and Celestino quickly ate a fried egg with manioc flour. We left the house through the side entrance and walked the sandy road to the beach. A beautiful full moon stood low above the town, and the fresh morning air enveloped our half-awake bodies. Casimiro had already arrived. "Good morning, Dutchman," he said, "ready to take a bath?" He was putting some tar in a hole in his canoe and enjoyed the apprehensive look on my face. The canoe could barely hold three people but I had insisted on going along. Some canoes had already left at dawn and could be seen tacking about in the distance. The two men moved the canoe to the surf in a zigzag motion, leaving a pattern in the sand that reminded me of the trail of a giant sea turtle before it buries its eggs on the beach. Casimiro and Celestino poled the canoe across the surf as I held on to the boards. They put the lateen sail, the jib, and the leeboard in place, and Celestino lit a cigarette. The sound of the surf ebbed away as our distance from the coast increased and our bodies adjusted to the swell of the ocean.

The decision about where to go depended on the direction and force of the winds and sea currents, the phase of the moon, and especially clues from yesterday's catches. Casimiro's cousin had caught some snapper yesterday near

Pedra Grande, and this gave the men, as they called it, their destiny (*o destino*). The coastline looked stunning. Camurim was embossed in palm trees and the beach made a thin golden frame below the sun-lit town. The emerald green sea gave an ever-moving base to this visual spectacle as we sailed north. When we saw the surf breaking on Pedra Grande, Celestino dropped the killick and the men slid shrimp on the hooks of their fishing lines. They had not been able to buy bait at the beach seine or from the shrimp trawlers, so yesterday they caught freshwater shrimp in a dead meander of the river, dragging a double stick net through the waist-high water.

What surprised me most during this trip was my complete lack of a sense of time. I asked Casimiro how he knew how much time had passed since our departure. "By the work done," he answered. I raised my eyebrows at Celestino. "Well," Celestino began, "first, there is the sun, but when the day is cloudy we look at the position of the trawlers. They take about one hour to complete their course. The movements ashore, like the seven o'clock bus and the Church bells, are also good signs, and the height of the surf on the shoreline is another. And finally . . . ," he paused, and dramatically rolled back his sleeve, "there is my wrist watch!" The time spent fishing at sea depended entirely on how soon we would catch fish. Fish will only remain fresh for about five hours and fishermen are sometimes forced to return home while the fishing is still good. We returned after nine hours at sea with the storage basket full of fish.

The earth turned under my feet as I stepped on the moist beach, so used had I gotten to the rocking motion of the canoe. Casimiro and Celestino gave me a large catfish and took a few others for their evening meal. The rest of the catch was strung into bundles and given to a teenage boy to hawk in the streets of Camurim. The men turned the canoe upside down to dry, and went home to take a bath and have a meal. Later in the afternoon, they rested a bit on the beach, and observed the catches of their colleagues.

Casimiro joined a group of friends to play soccer and Celestino decided to mend his nets.

### *The Composition of Fishing Crews*

Natalício and Amerino parted in anger after a conflict about the catch division. Natalício had loaned his canoe to his father's brother when Amerino, his younger brother, had bought a new canoe. The partnership seemed perfect because Amerino did not own any fishing gear while his brother's fine-mesh nets could feed both households. The two men, however, did not have any luck at fishing and Amerino decided to collect crabs in the mangroves after their return ashore. Several times he failed to show up to repair the gill nets, ignoring his brother's reprimands. Natalício could not mend the torn nets himself and began to deduct money from his brother's share to pay for a netman. Amerino replied that he understood this decision but as he was not a member of a corporate group, he was obliged to place the responsibilities as his family's provider ahead of his obligations as a fishing partner. His reduced income made him eventually break up with Natalício and borrow nets from a boat owner on a more favorable share basis. The elder brother asked his uncle to return the canoe and proceeded to fish alone.

Cassimiro commented about this incident that both men were at fault. "Amerino should have helped to mend the nets while trying to persuade Natalício to give him a larger share of the catch to support his poor family." He added that this incident further demonstrates that it is not wise to work with blood relatives because a father, uncle, or elder brother might assume that his seniority and kinship status give him more authority in economic matters. Work relationships between nonrelatives are more contractual and call for a greater respect between partners. Only four of the seventeen crews are composed of kin: two brothers, two half-brothers, two brothers-in-law, and a father and son. In these cases, there are no conflicts, because three of the four fishing partners are below twenty years of age and are eager to learn from their senior relatives.

Seventeen canoes are manned by two fishermen, while sixteen other canoe owners usually fish by themselves. The stability of these crews fluctuates with the seasons. Crews are most stable in summer when the fishermen use gill nets. The physical effort of retrieving fish twice a day and moving nets to other locations is so great that even canoe owners who like to fish alone sometimes ask apprentice fishermen to help them. However, a stable partnership is preferable, because most fishermen who do not have canoes own gill nets. The canoe owner and his companion can therefore cast the nets together in more locations than one person could by himself. Due to the joint effort, the men can quickly repair torn nets and still keep more than ten nets in the water, thus raising their profit and productivity.

This cooperation most likely comes to an end in winter, when handlining becomes the predominant fishing method. In April, net catches begin to diminish because the fish move away from the turbulent coast at the advent of the rainy season. Fishermen change from netting pelagic species to handlining demersal species of fish that reside near coral reefs and rocky bottoms several kilometers offshore. The distance to the fishing grounds and inactivity on days with bad weather may lower the income considerably and persuade canoe fishermen to switch to boat fishing. Canoe owners prefer to continue handlining in shallow waters, which can be done singlehanded. Most men who go boat fishing in winter switch back to canoe fishing in July and August to fish for snooks. Snooks appear unexpectedly, but they fetch such high prices that fishermen endure weeks of poor earnings to make one good catch of a dozen snooks.

Besides economic and ecological motives, personality conflicts and interpersonal tensions might separate crews. When a canoe owner, in the opinion of his partner, often arrives too late on the beach, or when a fisherman repeatedly contests the owner's decisions on where to fish, then the working relation might end because of mutual annoyance.

The main reasons for the overall stability of crews are a general agreement about the rights and obligations between canoe owners and their partners, and a mutual interest in each other's domestic situation. Canoe fishermen do not define their relationships with canoe owners in terms of a structural conflict between capital and labor. The importance of canoe ownership is downplayed and canoe fishermen—partly because of their possession of gill nets and the low price of used canoes—regard themselves as just as independent as canoe owners. In fact, canoe owners do not receive any portion of the catch as a return for this capital good. The only difference between owners and fishermen is the greater commitment of canoe owners to coastal fishing. They invested in dugouts to have the freedom to fish all year around, choose their own strategies, and work with whomever they feel comfortable as partners.

Crew members are either young adults who are still enlarging their stock of nets or fishermen who like to switch between fishing modes. Both appreciate the canoe owners' knowledge of the coastal fishing grounds and therefore consent to their greater influence on production-related decisions. They allow the owners to organize the work schedule, establish the proper catch division, and they help to mend damaged nets. The harmony of the work relationship between canoe owners and canoe fishermen owes much to their similar consumption demands and their genuine concern in satisfying those common needs together. They do not want to maximize earnings at the other's expense, nor are they driven by an unbridled urge toward capital accumulation, as is evident from the flexible, informal way in which catches are divided.

## *The Catch Division*

The two main forms of catch division are the joint line (*linha junta*) and the separate line (*linha separada*) division. A canoe owner and a fisherman who go handlining may either pool their individual catches—regardless of value or size—and divide the total in two equal parts, or they may keep the catches separate and sell their own shares. Canoe fishermen never prearrange the catch division but decide according to the outcome of the fishing trip. They use a separate line division when both make good catches, while bad catches are lumped together. If one fisherman catches a lot of fish but the other not, then the first will keep his catch and give some fish to his unfortunate partner. The costs of bait are always equally shared. The canoe owner does not receive any compensation for his canoe, although he may ask his partner to help him pay for minor repairs or a can of paint.

The same catch division principles are applied when both the canoe owner and his partner own gill nets. A more complex situation arises when one person owns all of the nets. This is the case with four crews. The fishing partners are young fishermen who have not yet been able to buy nets. Most net owners divide a poor catch in two equal parts, and a good catch in three: one part for the nets, one for the owner, and one for the fisherman. The fraternal conflict between Natalício and Amerino resulted from Natalício dividing poor catches in three parts, and later further reducing Amerino's share to a quarter, allotting half to the nets as the return-on-capital.

Canoe owners and canoe fishermen, due to their status as independent producers, have little credit in local stores and cannot rely on institutional arrangements which offer financial support during hard economic times. The allocation of the proceeds of fishing reflects a collective responsibility for the reproduction of the household. Canoe owners, who in general are a bit better off than canoe fishermen, reconcile their household demands with the needs of their fishing partner and fellow corporate group members. They take the needs of the fishermen into consideration and will suspend formal principles of catch division that harm the subsistence of less fortunate families. The production mode can operate properly without this support system, as is shown by the nonaligned canoe owners, but for most men redistribution and empathy for the wellbeing of each other's household are intrinsic to economic practice.

## *The Street Peddling of Fish*

Boys have been selling fish in the streets of Camurim for as long as people can remember. Canoe fishermen confess that they do not peddle fish out of fear of the evil eye (*olho grande*) of jealous colleagues (cf. Maloney 1976:131). Harm from envious glances might make them fall ill or have an adverse effect on their catches. Although this belief should not be underestimated and appears in other economic practices, it would be difficult for fishermen to hawk fish. The men need the precious time between and after short but strenuous fishing trips to relax and repair nets. The merchandise must be sold quickly to avoid spoilage and to take advantage of the not-yet-saturated market before other fishermen arrive ashore (Acheson 1981:281–284). The beach is the only place where fishermen sell fish directly to consumers. Prices tend to be slightly lower and allow consumers to have a first choice. The hours spent waiting for canoes which might not bring any fish at all discourage many people from buying fish at the beach.

Fish hawkers are eight- to fourteen-year-old boys who are often related to the fishermen as nephews or as the sons of neighbors or ritual co-parents (*compadres*). Francisco said that they are more trustworthy than unrelated teenage dealers, who sometimes pocket part of the earnings or fail to show up when the canoes arrive late. Surprisingly, not one fisherman's son of this age group sells fish. Murílio explained that he does not need the extra income from the labor of his children but prefers them to study so that they can find good employment ashore.

Hawkers earn a ten percent commission on the sales and sometimes receive fish that were left unsold. Francisco always gives 15 percent of the retail value to secure the services of his dealer permanently. Boys are known to leave their regular supplier for another who arrives earlier with a good catch, while during days with poor catches and rainy weather few street vendors are available. The earnings of a dedicated fish hawker can be substantial. For example, during the one-month period from January 10 through February 9 of 1983, Francisco and Murílio each made Cr$105,000 ($210). Their eleven-year-old dealer received Cr$25,000 ($50), which is more than the minimum wage of Cr$23,000 ($46) earned by most salaried employees.

After the fishermen beach their canoe, they divide the catch into small heaps according to the species. They do not weigh the catch but make bundles of fish (*rodas de peixe*) of a size and price that have proven to be the most marketable. A typical bundle weighs $1\frac{1}{2}$ kilograms, has three fish and costs Cr$500 ($1). Large fish are sold by the piece. Tiny, low-quality fish are strung into bundles of a dozen or more fish that sell for Cr$250 (0.50), or less. Although these prices may remain stable for several months, the amount of fish per fixed-price bundle decreased slowly during the 1982–83 fieldwork period to compensate for Brazil's triple-digit inflation rate. The decision about when to make a small decrement in weight is usually taken after a fisherman has experienced some increases in the price of basic household goods. He will tie together four fish that are a bit smaller; a month later he makes a bundle of three medium-size fish that gives the impression of being of similar weight as the previous bundle but is slightly lighter; then he joins one large and two smaller fish, and so on.

A sudden price jump of 50 percent or 100 percent for a typical bundle always follows upon a federally mandated price rise of basic foodstuffs such as bread, rice, beans, or meat. To somewhat neutralize the psychological shock to the consumers of a substantial price change, the canoe fisherman enlarges the bundle of fish while still raising the price per kilogram. For example, three medium-size fish that cost Cr$500 ($1) will be substituted for a bundle with five fish for Cr$1,000 ($2). After several weeks, one canoe owner begins to complain that the price of soybean oil has gone up and that the bread is becoming smaller. He reduces the total weight of the average bundle, and instructs his street vendor to insist on the price and tell suspicious consumers that everything else in town has become more expensive.

The hawkers carry the fish on wooden poles through Camurim after receiving retail instructions. There is no need to advertise the merchandise vocally because people can see the hawkers pass by in the street. The prospective buyer inspects the fish, inquires after the price, and invariably makes a lower counter offer. If there are few fish for sale in town or if the dealer has not yet completed his usual trajectory, he will reject the bid. Otherwise, he will negotiate. As a bargaining strategy, it is common practice for a hawker to ask for a higher price than the fisherman had given him. A higher price also raises his commission and gives him the opportunity to keep the difference without telling the fisherman. Normally, the vendor has very little freedom to haggle because he is not a middleman but acts as the fisherman's agent. He cannot divert too much from the preestablished price without asking permission.

A successful hawker must be a skilled haggler and also be aware of the town's socio-spatial division. A smart street vendor will personally offer expensive fish to affluent townspeople near the main street and spend considerable time praising their delightful flavor, while just passing quickly through poor neighborhoods with cheap fish, knowing that the consumers will approach him. Surprisingly, the neighborhood of summer residences at the northern perimeter of Camurim is often skipped because it is not a good market. Many wealthy vacationers employ local women to cook their meals and lay their hands on the best fish, while tourists who rent small houses in Camurim for a few weeks are more likely to buy fish at the beach.

The fierce competition for a limited supply makes the tourist season from December till March the period with the largest price jumps, although most canoe fishermen sell poor quality fish for a low price to local people who cannot compete with the greater purchasing power of tourists. The market expands rapidly but the supply of fish in Camurim remains the same because the produce of boats is sold to wholesalers, who take the fish to more profitable urban markets in Rio de Janeiro and Vitória.

## *Social Gatherings and Joking Relations*

Canoe fishermen spend much time together mending nets. The regular maintenance of fishing gear is essential for the optimization of the capital goods and at the same time provides an opportunity for strengthening corporate bonds. The netmending party is the most important social form for fishermen to enact their status as members of corporate groups. Fishermen seldom own enough gill nets to rotate them, so they only retrieve nets when catches decline because of damage or debris. Most of the damage is done by small sharks that gnaw their way through the nets and by crabs that feed on entangled fish. During summer, owners of wide-mesh caçoeiras repair their nets two or three times a week, four hours a day, while tainheira owners have to spend about five hours a day mending the more fragile fine-mesh nets.

On rainy days, the men hang the nets under a veranda at the back of their houses, but on hot summer days, each corporate group assembles in the shade of some coconut palms in their territory to enjoy the gentle sea breeze. The group seeks each other's company and sometimes helps to repair a colleague's badly torn nets. Usually, however, the turma gathers in front of the house of one of its most respected members, like Francisco, Camurim's most productive canoe owner. Francisco and his half-brother Murílio sit in a semicircle around the net that is tied to the doorpost and fans out across the sidewalk. Other fishermen, teenagers, and a passerby lean against the wall.

The first distinctive feature of these gatherings is the directional gaze of the participants. All visual attention is fixed on the hands tying the knots. Even bystanders who talk to one another do not divert their eyes, mesmerized by the swift movements of the menders. Francisco and Murílio are the center of attention because they uphold an activity that for its duration gives the meeting its focus.[11]

The second characteristic is that the menders usually direct the conversation. The netmending party constitutes a speech event, organized through tacit rhetorical agreements (Hymes 1972). Francisco and Murílio generate the meeting, which gives them the right to define the conditions of social interaction and impose their standards on the group.

> "This morning I heard the catfish singing," started Murílio. "This humming sound must be their mating call. It's that time of year again." He smiled with an indecent expression on his face. "I think that the coast is still too rough for the fish to spawn," replied Francisco. Valderí, his 16-year old nephew, noticed, "yesterday a canoe owner from the turma do rio placed his gill nets near the river mouth and he is expecting the catfish to come to shore any day now, and . . . " Valderí halted because Casimiro put his hand on Francisco's shoulder and whispered, "Did you know that

his wife is having an affair with the undertaker?" Francisco looked at him annoyed and redirected the conversation to their ongoing territorial dispute with the turma do rio.

The menders seem to ignore all inopportune remarks that do not deal directly with fishing. Francisco likes to talk about soccer and nobody can stop him when his voice cracks in excitement, but he and Murílio are the only participants to make such digressions. The other men respond but, after a brief silence, return to economic matters. Bystanders may tell humorous stories unrelated to fishing as long as these are cast in a speech pattern that clearly sets the stories apart from the flow of conversation. Once the men put away their nets, the setting transforms into an unfocused gathering. Now, most participants may contribute to the conversation with varying subjects.[12]

Joking relationships are the third characteristic of netmending parties and especially streetcorner gatherings. Canoe fishermen make a distinction between *divertimento* (diversion or amusement) and *brincadeira* (mockery or teasing). Divertimento is diversion for the sake of enjoyment, as in playing cards or telling an anecdote. Brincadeira is playful joking at the expense of another without, however, intending any malice. A person will reveal the intention of his joke as brincadeira to his audience by pouting his lips in the direction of the victim. This pouting is not seen by the victim, but the joker is indicating to others that his seemingly serious remark is only made in jest.

Brincadeira is common between canoe fishermen and the teenagers who only fish occasionally but who will eventually become professional fishermen. The asymmetrical joking relationship with its functional implications, as described by Radcliffe-Brown (1952:90–100) in terms of both detachment and attachment, seems to apply here. Murílio likes to tease Valderí about the contrast between his dark skin and his smooth hair which he bathes weekly in coconut oil to make it soft. He pulls at his clothes, gently kicks and slaps him, and throws orange peels on his head. Valderí retaliates in similar though milder ways until Murílio takes him in a forceful grip and challenges him to strike back. Amidst the laughter of the rest of the group, the teenager squeals in alleged pain and is finally released. Such encounters happen several times a day and are a common diversion for the group.

The joking, if we accept Radcliffe-Brown's interpretation, is the result of the ambiguous relationship between adults and teenagers. The two are always in each other's presence even though the teenagers are not yet members of the corporate groups. In the near future, however, they are going to have the same status as the current fishermen. Joking seems a means of social control to socialize aspirant fishermen into the values of the corporate group, to make them assume the social identity of canoe fishermen, to reduce their adolescent self-centeredness, and to broaden their loyalty to parents, siblings, and relatives with a similar concern for their future colleagues. Some attention to the household must be diverted to the corporate group. They must learn to be tolerant, helpful and acquire a sense of responsibility for other turma members. The aspirant fishermen must be instilled with the values of hard work and sharing that will continue the redistributive obligation that benefits all fishermen. Teenagers are teased for being lazy, of living on other people's backs and are told that the time has come to grow up and assume the responsibilities of life. Once teenagers have been fully initiated into the standards of the turma and begin to fish permanently, the teasing ceases (cf. Howell 1973). Joking relationships thus seem to be forms of social control in function of the social structure of the corporate groups.

A different interpretation appears when the weight of the analysis is shifted from the structural form of the joking relationship to its content. The most common teasing remark, and the one that elicits the most emotional reaction, is to accuse a teenager of being a masturbator (*bombeiro*, literal meaning: fireman). The anger of the youth at hearing this slur and the consecutive laughter of the crowd derive not merely from the shame of his alleged conduct but from a chain of binary associations of metaphors which show that the teenager is excluded from the corporate group because of his immaturity. These associations define the conditions of the corporate group through the contrasts of predicative metaphors. The bombeiro stands in opposition to the *homem*, the male adult. The homem helps to generate children through his procreative capacity and thus biologically reproduces society, while the bombeiro wastes his semen in an individual, self-gratifying manner. All fishermen are heads of household carrying responsibilities for wives and children. The bombeiro is a *malandro*, a rascal, who manipulates his social environment for his own benefit without contributing any labor to its maintenance (cf. DaMatta 1981:194–235). The homem is a *trabalhador*, a worker, who objectifies his actions into goods that are used to sustain the household, the corporate group and, ultimately, society.[13]

The analogies between teenagers and canoe fishermen are so striking that bombeiros have emulated sea tenure into a form of beach territoriality. They have divided the beaches frequented by bikini-clad and sometimes topless tourists into a number of zones that are each "guarded" (*vigiado*) by several bombeiros who have the exclusive right to daydream about sexual affairs with the women in their area. This beach tenure does not of course have the characteristics of sea territoriality, nor are the zones defended with vigor but only in jest. Still, the imitation of the corporate resource control is significant to understand the joking relationship both as a symbolic challenge to the dominant corporate relations of the fishermen (Douglas 1968), and as a way of identification with the fishermen.

On several levels of meaning, the bombeiro consumes, while the homem produces. Masturbation relates to procreation, as individual relates to society, as consumption relates to production, and as leisure relates to labor. Fishermen produce what they consume and, in addition, they satisfy the demands of several dependents. Teenagers are of an age to sustain at least themselves, but they still sponge on the labor of others. The metaphors express and define the identity of economically active males and indicate the significant cognitive categories of social organization (Fernandez 1974). The binary oppositions reveal the practices that contribute most to the social identity of the men and the teenagers. They show a central preoccupation with work, family, social responsibility, duty, and dependability. The binary opposition of these metaphors gives a glimpse of the interpretations of the economic practices of canoe fishermen, which is elaborated in another way by the discursive conflicts about the canoe as central to the integrity of their personal and social identity.

## *Canoes and Independence*

Canoe fishing has an economic organization in which most operational decisions are taken in Camurim. The use of marine resources is restricted to a small number of producers with a relatively simple technology and division of labor which make canoe fishing, in relation to boat fishing, a small-scale endeavor. Both production modes are encapsulated by the Brazilian economy but canoe fishing utilizes locally available capital goods and functions through personal economic relationships, while boat fishing is highly dependent on national and institutional forces that enhance the vulnerability of its organization. The canoe orients the canoe fishermen towards self-determination and autonomy because its small-scale operation cannot capitalize on the infrastructure that supports boat fishing.

The control of shallow water resources by corporate groups gives canoe fishermen some security over the stability of their production mode. Sea tenure is a safeguard against overfishing because access is limited to a group of people who through peer pressure may enforce the use of fishing methods that are benign to the marine habitat (Cordell 1978; Robben 1985). The corporate group also shares the responsibility for the welfare of its members through a system of redistribution that protects families from starvation and prevents fishermen from having to sell off their fishing gear when the catches are poor.

The capital goods most adequate for the coastal habitat are relatively cheap and widely available. Fishermen do not have to, and purposely do not want to, enter into business arrangements with patrons or banks to finance the purchase of canoes and gill nets. Zé Maria told me of his doubts about borrowing money to buy a canoe.

> Last year when I had to buy a canoe because I had broken up with Ricardo, I thought of asking Celso Andrade [the

> landowner] for money. But then I thought "I'm going to ask him for money, and then he will tell me that he doesn't have any, and I will lose face (*passar vergonha*)." Then I thought of going to the bank to get a loan, but for that you need a cosigner. I know enough people to guarantee my loan but to run after them like a dog to be my cosigner is not for me. Finally, I decided to borrow part of the money from my brother-in-law who is also a canoe fisherman, and I paid him off in six months.

Most canoe fishermen prefer to save money and slowly expand their capital instead of risking dependency upon powerful investors. The investment necessary to become an independent producer can be obtained in a couple of years. It thus greatly reduces the possibility for a rapid concentration of capital and the development of a stratified mode of production. Eight of the fifty active canoe fishermen do not own nets, mainly because they began fishing recently. Four of these young men, however, cast and repair nets owned by retired fishermen. The forty-two net owners each have an average of seven gill nets. Out of thirty-three canoe owners, only one person owns two canoes, a small one for handlining and a large one for casting nets with a fishing companion. This capital distribution reduces the variation of income among the fishermen and implies only small differences in living standard (income figures of canoe owners and canoe fishermen will be compared in chapter 5). Finally, the teenage hawkers who sell the products to the consumers are intermediaries with little power. They do not act as middlemen who manipulate the consumer prices and tie the producers to them through credit arrangements. Canoe fishermen have a firm grip on the fish sales and know that a high local demand for fresh products allows them to take full control and advantage of the market.

The accessibility of capital and natural resources impedes the development of social classes in the canoe fishing mode and thus ensures the independence of the producers. This "freedom and independence" (*liberdade e independência*), however, must not be taken for granted. Overconfidence can quickly lead to a deterioration of capital goods and a loss of autonomy. Older canoe fishermen repeatedly press younger colleagues to maintain their nets in good condition "because they guarantee your independence." Older men have experienced the dependence on, and subservience to, boat owners when they temporarily switched to boat fishing. They know that canoe fishing demands much motivation to fish seven days a week, to row for kilometers on end against strong winds, and to confront a turbulent and dangerous sea to visit three or four nets that might be empty. The small daily yields force canoe fishermen to work incessantly. However, this independence is not only threatened by a slackened motivation to maintain the capital goods, but it is also assailed by a mode of production which tries to expand at the expense of canoe fishing. Their independence as producers and, at the same time, their cooperation through corporate groups give them the "endurance" (*resistência*) to be independent. Francisco elaborated on his drive to work hard, "I do not want to debase myself, to subject myself to the rich, the landowners, or the boat owners. They only want to step with their feet on a *caboclo*" (poor person, literal meaning: mestizo).

In the absence of credit in local stores, only work will yield the money for household purchases. The corporate group provides some social security but its redistribution system is based on reciprocity. Fishermen who persistently ask for fish but never return these gifts will inevitably fall into disfavor and be excluded from the allocation service. Corporate pressure, incessant consumption demands from the household, and the quest for social esteem in the peer group make canoe fishermen work hard and, as a result, makes them maintain their independence.

Nonaligned canoe owners argue that the members of corporate groups have renounced their autonomy as producers and lost their social identity as independent men. They interpret sea tenure as dependence while corporate groups interpret their cooperation as a means of securing their independence under the rising pressure on the marine resources by an expansionist mode of production. What are the justifications for these opposing interpretations?

Nonaligned owners accuse corporate groups of restricting the movements of the canoe fishermen and dampening their incentive to work hard. A corporate member feels less inclined to be aggressive at sea, to move his nets frequently to different locations, and to use his knowledge and skills to discover new fishing grounds. Protected by the limited entry, he will leave his nets stationary for longer periods, consider other zones off-limits, and will thus impede the mobility of more enterprising fishermen who feel obliged to respect sea territoriality. The nonaligned canoe owners charge furthermore that limited entry is more conducive to overfishing because the members of each corporate group are confined to one area and will continue to cast their nets even though major schools of fish have migrated elsewhere. Instead of following the schools in their movements along the coast, they will exploit the seasonally poor resources to exhaustion. Nonaligned canoe owners who do not accept the alleged scarcity of marine resources, but attribute the declining catches to supernatural wrath, feel only strengthened in their resistence to sea tenure.

The proponents of sea tenure do not necessarily deny the abundance of fish or diminish the importance of making ritual offerings to the Sea Goddess, but they are convinced that the limited operation radius of the sail-driven canoes inhabitants them from chasing the evermoving schools across the vast shallow waters of Camurim. The division of the sea into fishing spaces with limited entry allows each corporate group to optimize its output without all of the canoe fishermen crowding the fishing grounds closest to Camurim. Ecologically, they defend their position by arguing that the redistribution of fish in times of need has not only improved their social welfare but has also protected the marine environment from depletion by their abandonment of fine-mesh nets (*tainheiras*).

Nonaligned canoe owners counter that the corporate fishermen with their wide-mesh nets (*caçoeiras*) will not be able to survive independently if the corporate group should disintegrate. Quarrels might develop, the price of high-quality fish could drop, and declining catches might place too much strain on the redistributive system. Without the support of the corporate group, the owner of wide-mesh nets could not survive on the infrequent catches of large fish. Nonaligned fishermen sense that their colleagues have sacrificed their autonomy and are compromising not just canoe fishing but a particular way of life.

Corporate members continue to be convinced that only unity against the overwhelming force of trawlers and netfishing boats can safeguard their future by fairly dividing at least those shallow waters that cannot be exploited by motorized vessels. They argue that the personal identity of the nonaligned canoe owners, their self-identification with the canoe fishermen of the past, is at odds with reality. Canoe fishing, which already was enveloped in a national market economy, has now also become part of a pluriform fishery dominated by an expansionist mode of production. Both fishing modes are to a greater or lesser extent dependent on market exchange, but their principal difference lies in the control over the production process by the fishermen who supply the labor.[14] Corporate members acknowledge that they have adapted to these new economic circumstances and developed in sea tenure an organized form of resistance but insist that their social identity as independent and autonomous canoe fishermen has not fundamentally changed.

# 4
# Boats on the High Seas

"When Felipe left Camurim to go fishing, he met the boat as it returned from the sea," so recalled Elmício the dramatic events of the mysterious death of Francisco. "As the boat arrived I was happy to see it," continued Elmício. "We had had some days of bad weather and I am always concerned about the safety of the men at sea. I asked the captain how things had been. And when he told me that Francisco was missing, my legs began to tremble. And I said to myself, 'Thank God.' Because Francisco was the only fisherman in the crew who didn't have a family, you know. While the investigation was under way, I could not fish with the boat for seven months." "Why so, Elmício?," I asked. "Because Francisco was the only crew member who did not have documents." He resumed his narrative. "When the six crew members arrived they told me what had happened. Francisco had strung together some triggerfish and placed them on the boat's bulwark. The fish slipped into the water and he went after them. Holding on to the fish, he drifted away from the boat with the current. Zé Pequeno jumped after him with a rope. As the two were being carried off further away from the boat, Zé Pequeno told Francisco to let go of the fish and swim back to the boat, but he didn't want to. As Zé swam in a southern direction so that the sailboat could pick him up, Francisco was being dragged further and further away. The body was never found."

> Did Francisco really drown? Or was he murdered? Many people feel that the truth was never revealed because the crew made a secret pact to conceal the real course of events. The informal investigation by a corrupt lieutenant of the Navy was quickly closed, even though eyewitnesses said that the crew members were very nervous during the inquiry, and refused to talk about the incident afterwards.[1] Francisco's half-brother Mario gave me his account. "My mother never received any damages for the loss of her son because Elmício had paid off the lieutenant. Elmício had the whole thing covered up since, with the exception of Francisco, the other crew members were unregistered fishermen. When my mother went to him to receive my brother's belongings and the money he had earned on his final fishing trip," said Mario, "the boat owner insisted on deducting the food expenses of the trip. When a person dies, the bank cancels the rest of one's debts. But not Elmício. He had to take money even from a dead man. Now, he is paying for his crimes. He has been having bad luck with his boats because God wants him to suffer. God will make him pay the final bill. He will be weighed, and will have to fly around in space without any destination. He will be in hell."

What is remarkable about this story is not so much the mystery surrounding Francisco's death or even the suspicious silence of the crew, but the deep impression left in Camurim of the absentee owner's conduct during this drama. The reproachable attitude of Elmício towards the mother of Francisco is often mentioned by boat fishermen as illustrative of their exploitation by boat owners.

Many discursive conflicts between boat owners and boat fishermen revolve around their personal relations and the structuration of economic practice. Their discourse is replete with implicit references to their different interpretations of agency and structure. Boat fishermen regard the economic organization as an interpersonal universe of producers, capital owners, dealers, and consumers. The economy operates through friendship, trust, obligations of reciprocity, personal favors, graft, favoritism, manipulation, and other features of social relationships. This interactional conception of the economy makes them only seldom put the blame for their personal lot on the larger economic institutions of the Brazilian economy but instead on particular individuals, mainly the boat owners. The boat fishermen place the boat owners at the center of their economic universe—powerful, resourceful, influential, and dominant.

Boat owners regard their small fishing enterprises as part of a structural framework of banks, federal agencies, communication systems, markets, and large corporations. They analyze the economy with concepts such as inflation, glut, interest rates, and supply and demand. They are aware of the importance of world oil prices, federal development programs, loan conditions, and related aspects of the economy. They find themselves at the mercy of a conglomerate of national and international institutional forces, and believe that the urban fish dealers at the same time embody and manipulate impersonal market forces.

"Acting from different perspectives, and with different social powers of objectifying their respective interpretations, people come to different conclusions and societies work out different consensuses" (Sahlins 1985:x). However, a working consensus between boat owners and boat fishermen does not imply that one can therefore isolate a folk model of structuration which ultimately guides their actions. Conflicting interpretations of the structuration of economic practice lead boat owners and boat fishermen to different conceptions of the relation between economy and society, and entail different strategies and plans of action to realize their interests and aspirations. Boat owners and boat fishermen reproduce those conflicting interpretations in their economic practices by choosing respectively institutional and interactional avenues of social mobility, prestige, and capital ownership. Unintentionally, they transsubstantiate those different conceptions of economy into structures that sustain their contradictions within the whole of economic practice.

The interactional interpretation of economy makes it difficult for boat fishermen to draw a sharp distinction between the economy and other social domains, because for them there is not a clear break between their social relations as fishermen, as members of the community, and as the providers of their families. As was explained in chapter 1, they experience fishing as an economic practice whose content transcends the production, distribution, and exchange of economic goods.

Boat owners argue that the boat fishermen confound roles and obligations which pertain to discrete social domains. Their institutional conception of economy allows them to confine economic relations to clearly bounded institutions. Their justification of refusing to change the distribution of the catch during a poor fishing season—as is common practice among the canoe fishermen—serves as an example. They argue that consumption must always be subordinate to production, even if certain basic demands have to be suspended temporarily, because their financial downfall will in the end be more harmful to the fishermen and their families than a brief period of hardship. Boat owners are afraid that exceptions to their inflexible position will erode the boundaries set by the conditions of the economic relationships, provoke incursions from other social domains, and entangle them in the households of fishermen. They can pretend to be concerned only with issues of management, and can condemn fishermen for the introduction of extraneous demands because a successful enterprise is the best asset for realizing their personal objectives. The institutional economy has an intrinsic value for the boat owners because their social identity as capital owners makes them accumulate power and prestige in other domains of society.

These enduring quarrels not only bring out the different interpretations of the economy between boat owners and boat fishermen but the "relationships themselves are put at issue" (Sahlins 1981:72). Questions are raised about hidden motivations, acts of compassion are examined with suspicion, and advice is mistrusted. An awareness emerges about how profoundly hierarchical the relationship between boat owner and boat fisherman is, and how this reciprocal yet unequal dependence is a social consequence of the means and mode of production.

The interpretations of the dependence entailed by boat fishing make the boat owners and boat fishermen realize how their lives have been affected by the introduction of boats. I will analyze the boat as a cultural focus through a description of how aspiring boat owners acquire capital, wholesalers control the marketing of fish, and how the organization aboard ship and the catch division reflect the social hierarchy of the crew. Although boat owners and boat fishermen do not perceive the same social consequences of the boat, they share a concern for the social inequality characteristic of boat fishing, the chances of status climbing it offers, and the various ways in which it relates the economy to other realms of society.

## *Finance and Boat Ownership*

Elmício still gets angry when he talks about his initial involvement in fishing. "Years ago, I had a store and lots of cattle. I owned so much land near the estuary of Rio Camurim that people called me Elmício da Barra." He paused. "I should have never allowed myself to be drawn into this mess. Well, in the mid-1960s, some men began to motorize the large canoes that had been used for the transportation of cocoa. The fishermen got to the fishing grounds more quickly, but they couldn't stay overnight and the costs were too high. In 1969, Valdomiro bought a sailboat in Pôrto Seguro. Valdomiro's crew could remain at sea for several days and bring in hundreds of kilograms of salted fish. His greatest preoccupation was marketing the catch because the canoes sold fresh fish to the people here, while the business from itinerant dealers who roamed the interior on their mules was too uncertain. Valdomiro asked for my help because I had many business contacts. I succeeded in exchanging the fish for groceries with a dealer in Serrania. But, then, this s.o.b. tried to cut me out of my own business. He became jealous and abused my friendship. Then and there, I swore to restore my honor and break Valdomiro, even if I would have to sell all my cattle. I bought two eleven-meter sailboats—I still own them—and again began to sell fish to my contact in Serrania. Valdomiro was obliged to sell his boat and today he's just a gardener at the bank."

In 1970, Elmício motorized the two boats. Profits were so high that in the early 1970s the public notary of Camurim bought three eight-meter open-hull motorboats. In the following years, two more boats were anchored in the harbor. One was bought by a baker and the other belonged to a boat owner from another town. The fleet expanded from seven to seventeen in 1975 when an entrepreneur from the city of Vitória arrived with ten boats to found a major fishing enterprise in Camurim. He left at the end of the 1970s because of personal problems at home. Five boats were sold to their captains and the rest were taken back to Vitória.

Till the end of the 1970s, none of the boat owners were former fishermen. Canoe fishermen did not have enough savings or property for sale that could finance the purchase of motorboats. This impediment on capital investment was removed when the fishing entrepreneur from Vitória and the public notary from Camurim decided to sell their boats. They learned of the possibility of obtaining loans at the Banco do Brasil and introduced their best boat fishermen to this new source of capital. The loans are still available and have been the greatest incentive for the expansion of the fishing fleet from twelve boats in 1979 to thirty-two boats in 1983. Bank loans have allowed many boat fishermen to become boat owners, which has led to the dissemination of capital goods. In 1983, the grocer owned five boats, three owners had two boats each, and twenty-one men each owned one boat. Aside from Elmício, only five men were not former fishermen, each owning one boat.

The local branch of the Banco do Brasil offers loans with an annual interest rate of 12 percent. This loan is subsidized by the PROTERRA, a federal incentive program for fishing and agriculture that has financed many cocoa planters in the region. The interest rate is extremely low in view of an accumulated annual inflation rate of 211 percent in 1983. Loans are to be paid off in five to eight years.[2]

The demand for loans remains great. Unfortunately, the strict qualifying conditions limit the number of suitable applicants mainly to those men who already own boats or some other property with which they can underwrite their own loans. Any other applicant needs three references and a cosigner. The references are provided by colleagues who declare that the candidate is a respected and skilled fisherman who will run a successful fishing enterprise. The most important but also the most difficult requirement to fulfill is to present a cosigner. The cosigner (*avalista*) is ultimately responsible for the payment of the loan. The bank wants a person with ample resources such as real estate, plantations, forests, cattle, or boats that can be mortgaged or appropriated if necessary. Although not one boat owner has ever defaulted on his loan payments—especially because of the low interest rate and the flexible arrangements for paying off arrears—several local cosigners have lost large amounts of money to insolvent borrowers in agriculture. These bad experiences have made wealthy townsmen shrink for committing themselves to a fisherman who has no assets beyond his good faith.

The need for a cosigner is the single most important institutional bottleneck for economic mobility in the boat fishing mode, and therefore much of the social interaction ashore between boat owners, boat fishermen, and the town's elite turns around this issue. Boat owners strive to maintain their ties with cosigners who helped them in the past, and upwardly mobile captains and fishermen try to cultivate a network of wealthy townspeople who in the future might assist them in securing bank loans.

The increasing difficulties in obtaining investment capital have given the financial institutions and the landed elite of Camurim more power in the boat fishing mode than ever before. For owners, boats entail dependence (*dependência*) upon those in control of the financial sources. Not one boat owner can claim the independence so valued by canoe owners. The incessant drive toward technical innovation and the ensuing expectation of rising productivity make new loans always seem justifiable. Boat owners are aware of their perpetual indebtedness to banks and financial guarantors but, as Raimundo stated, "who doesn't owe, doesn't own" (Quem não deve, não tem). Independence equals poverty. It reminds them of the modest, frugal life-style of canoe fishermen, or of the low status of boat fishermen. They are willing to pay the price of indefinite dependence as long as they can reap esteem and prestige as the fruits of capital ownership. The boat is not purely a source of profit but a means to attain high public status, political influence, and identification with the town's elite.

Different interpretations of the economic organization guide boat owners and boat fishermen in problems of finance and investment. For boat fishermen, the avenue to status and success does not start in formal economic institutions. Since boat fishermen have an interactional conception of economic practice, they search for interpersonal solutions. They focus their time and effort on social events through which they hope to establish patronships with landowners and planters. They imitate the manners of boat owners, dress similarly, and try to be equally generous at informal gatherings. Those who believe these attempts to be of no avail will simply give up without exploring alternative routes and, in effect, will reproduce the interactional structure they resent.

This distrust of government institutions, development programs and commercial banks is historically well-founded. For centuries past, the landowning elite has personalized the public sector, using law, politics, government, and business as extensions of their social position while subjecting the underprivileged either to the impersonal bureaucratic operation of the social system or obliging them to reap minor benefits from patronage and subordination (DaMatta 1981:169–185; Schwartz 1973:171–190).

Boat owners suggest that the mobility channels through which they rose socially and economically are still open and sneer at the boat fishermen who point at the obstructions as being lazy, wasteful or poor by choice. They presuppose the availability of cosigners and thus assume that fishermen who do not borrow money are inept or without ambition. "There are fellows here who do not have capital because they do not want it," said Raimundo. "There is no problem in getting a bank loan because the local politicians will serve as cosigners to get votes. You only need their acquaintance (*conhecimento*) to have them help you."

However, boat owners who in conversations with the fishermen deny the existence of social impediments, and emphasize hard work as the only correct strategy, are aware of the importance of personal relationships in business. They propose solutions to which they themselves do not adhere, and employ various social strategies to circumvent or force open institutional channels. They befriend loan officers to influence their decisions, apply to federal incentive programs under false pretenses, borrow money in one bank at a subsidized interest rate and deposit it in another bank at a higher rate, and buy nets at a fixed price but purposely delay payment. They also use their power as capital owners to influence labor disputes, make boat fishermen pay for repairs, overcharge for running costs, refuse to pay for their health care dues, and withhold income temporarily to make crews more dependent. Boat owners manipulate and reinforce an institutional structure whose existence fishermen fail to recognize because of their interactional perspective.

## *Fish Dealers and the Urban Market*

The major institutional constraint on the overall development of the boat fishing industry is the marketing of fish, not the financing of the fleet. After the establishment of a trade relationship between Elmício and the dealer from Serrania, the next major marketing innovation was the purchase of an ice-making machine by the public notary in 1974. The machine made ice blocks which were placed in the boats' storage holds and allowed crews to return with frozen instead of salted fish. No valuable time at sea was lost cleaning and salting the catch, and frozen fish was in much greater demand. Although this new marketing strategy improved the vitality of the emerging fishing industry, ice production and the storage capacity of the freezer were too small to justify the purchase of more boats.

The arrival of the fishing entrepreneur from Vitória advanced the sale of fish from regional and municipal markets to national and urban commercial centers. A new, unlimited market opened up which could absorb any quantity of fish and finally relieved boat owners of their preoccupation with selling exceptionally large catches. The entrepreneur introduced refrigerated trucks which supplied the boats with cheap ice and abundant bait.

Refrigerated trucks are loaded with ice in Vitória and Rio de Janeiro and twice a week make the respectively 500 and 1,000 kilometer trips to Camurim. The last 40 to 50 kilometers are made over dirt roads that may turn into insurpassable mud tracks during the rainy season. Each truck carries boxes of salted sardines and eight to twelve tons of crushed ice. Without these two products, the fishing fleet cannot operate properly. Occasionally, a boat owner who is in conflict with the wholesalers returns to salting, but the crew does not like the work and salted fish is difficult to sell.

The need for bait is as acute as that for ice. In the past, the small number of boats were furnished with bait by local beach seines and trawlers. Fishing trips were short and fresh shrimp was excellent for catching shallow water species. Overfishing in the coastal waters forced boats to go to the edge of the continental shelf to fish on species that as bait preferred sardines to shrimp.[3] Sardines (*Sardinella aurita*) were delivered to Camurim by the wholesalers. This external dependency soon extended itself to oil, fuel, nylon line, hooks, nets, spare engine parts, and navigational equipment. With the exception of the last two, all other goods can be bought in Camurim but local prices are higher than those in the city. This dependency upon fish dealers is not necessarily detrimental to the parties involved but makes people in the fishing industry more vulnerable to uncontrollable events outside Camurim. A broken truck, mud roads, fuel price hikes, overexploitation of sardine stocks in the south, changed loan policies, or import restrictions on spare parts can severely damage the boat fishing industry.

Wholesalers consciously cultivate the dependency relationship with the boat owners. The dealers who visit Camurim work for large companies which buy fish in many harbors in the states of Bahia, Espírito Santo, and Rio de Janeiro. They know that they can ruin the fishing industry in Camurim without suffering any significant loss themselves. Paulo summed up the business practices of the urban dealers, "They divide the coast into areas so that they have the exclusive right to buy fish. They operate just like the bus companies which control their routes in order to avoid competition."

Wholesalers deny that they collude in the purchase of fish but admit that they seldom use aggressive business tactics to enlarge their market shares. Certain price developments in the city may induce a fish dealer to raise fish prices. Immediately, he will attract business because there are no formal contractual arrangements or long-term debt and credit relationships between boat owners and wholesalers. A price rise will quickly provoke a similar response from another dealer, which might lead to a counter offer, and so forth, until either one wholesaler decides to withdraw or both stop hiking their prices. The remaining dealer will only enjoy his single dealership briefly because soon another competitor will appear to occupy the vacated niche.

> At the beginning of May 1983, Cerezo from Vitória began to provide ice free of charge to his customers. His competitor from Rio de Janeiro could not follow suit because of higher transportation costs, but he slightly raised the fish prices. Cerezo responded with the same price increment and, at the end of the price war, dominated the entire business. He was giving away around 25 tons or Cr$187,500 ($375) worth of ice every week. Within weeks, another middleman from Rio de Janeiro entered the business offering the same conditions, and, in addition, he promised to pay for the catches immediately upon receiving them, while Cerezo still reimbursed boat owners only after selling the fish in Vitória.

The cyclical oscillations between single and multiple dealerships triggered by the price wars are beneficial for the fishing economy, but boat owners and boat fishermen do not have enough power to exert any real leverage. Price rises are not a direct reflection of fluctuations of supply and demand but come from the dealers' urge to increase their market share, and from allowing the boat owners to play them off against one another. Boat owners may threaten to sell their catches to another dealer if prices are not changed, or they may strike separate deals with both middlemen.[4]

Wholesalers attempt to maximize profits in several ways. Two truckloads cannot supply all the boats with ice, so the dealers give preferential treatment to those crews who generally have large, high quality catches. Successful crews can have as much ice as they want, they may select the bait, and are given more credit to cover operating costs.

Dealers favor liners over netfishing boats. Handlining brings species of fish of a better quality than the smaller, fishbone-ridden species caught with gill nets. The low quality fish are of the same species as those sold by canoe fishermen, but fetch much lower prices from wholesalers. Most dealers give boat owners permission to peddle less desirable fish in the streets. However, the quantities caught are too large to market locally and consumers prefer the fresh fish offered by canoe fishermen to the half-frozen, unappetizing-looking fish caught by boats. Furthermore, consumers are reluctant to buy from opportunistic, unknown peddlers with whom they cannot initiate an equilibrating trade relationship that guarantees the replacement of spoiled fish (cf. Plattner 1983).

Dealers make their greatest gain through the local unawareness of the current fish prices in the city. They are deliberately vague about their profits and claim that the supply and demand situation at the auctions makes prices fluctuate rapidly. Boat owners and boat fishermen are completely unaware of the high profits. They only ask for price rises when the running costs of fuel or bait go up, but such increases are always proportionally less than the corresponding price increments in the city. The following transaction gives some idea of profits of the firms from Rio de Janeiro and Vitória. I do not want to suggest that they never incur losses because the prices do collapse at times when bumper catches flood the market. Yet, their financial situation is generally better than they profess because they are not bound by contract to purchase fish. As soon as the fixed price drops below the market value, they will pull out of Camurim.

At the large fish market on Praça 15 de Novembro in Rio de Janeiro, the price of the locally favorite sea bass (*badejo, Acanthistus brasiliensis*) never fell below Cr$1,000 ($2) during 1983 and even reached Cr$3,500 ($7) per kilogram around Holy Week.[5] Wholesalers were buying sea bass in Camurim for Cr$300 ($0.60) in January 1983 and for Cr$500 ($1) six months later. During one transaction in June of 1983, a truck bought 3,500 kilograms of fish for Cr-$1,100,000 ($2,200) and paid Cr$150,000 ($300) on ice, fuel, and taxes.[6] The merchandise was sold for Cr$4,000,000 ($8,000) in Rio de Janeiro and thus yielded a handsome profit of

Cr$2,750,000 ($5,500)—of course excluding the depreciation of the truck, maintenance, sales commission at the auction, and the wage of the truck driver.[7]

When I showed this calculation to the dealer, he dismissed it as deceptive and mentioned his financial entanglements with the boat owners. "They demand material from outside—spare engine parts, fishing gear—and they want this material on credit. I buy all these things in Vitória and they pay it with fish. Yet, all this money is tied up and cannot be put to work. . . . The owners pay months later, and if I try to collect interest they will not pay. I have lost quite a lot of money with these transactions and I am going to cut them all. I do it only to keep their business, not to make any profit."

The marketing relationship between fish dealers and boat owners has a subtle sense of domination and submission. On the surface there is much congeniality, but behind the backs of the dealers, the boat owners complain about their relation as one of submission (*subordinação*), subjection (*sujeição*), and feigned humility (*humildade*). Boat owners have to concede to the commercial conditions dictated by the dealers who control the distribution channels with the sale of supplies and the purchase of fish. Boat owners must treat the dealers with more respect than they deserve, and still establish a mode of interaction that allows them to cancel their business deals when better offers appear.

This interpersonal submission is complemented institutionally. The boat owner feels that he has to surrender his small enterprise to economic forces over which he has no control and which are presented to him through abstractions such as price, supply and demand, glut, and scarcity. The boat owners express a sense of helplessness in the face of these erratic forces whose effects come to them through accounts given by the middlemen and through the evening news on television. The dealers use the alleged uncertainty of the market to suppress the fish prices and maintain them at a relatively low level. Still, boat owners prefer fixed to fluctuating fish prices because they do not receive any direct information about the price developments at the urban auctions. Fixed prices, low as they are, guarantee at least a calculable income and tend to reduce fraud "because the dealers cannot be trusted."[8] "Every fish dealer is rotten (*safado*)," Rodrigo assured me. "He never tells the truth. When you go to receive your money, he already says 'In a little while, in a little while,' but he never comes. You always have to run after him because he constantly misleads us, saying that he is going to bring the money, and then he says he doesn't have any."

Boat fishermen do not share the boat owners' concern for investment debts and deny the assertion that low fish prices are only the dealer's fault or the effects of market fluctuations. "If the fish prices are determined by the supply and demand as the boat owners say," so reasoned Chico, "then why doesn't the price change? It doesn't change because the dealers and the owners keep the price down. Look at the price of fish in the street. When there are a lot of fish the price goes down, and it goes up again when there are few fish."

Boat fishermen understand supply and demand as the negotiation between dealers and consumers similar to the fish peddling practices in Camurim. Greater demands implies higher earnings. Fixed prices do not serve to cushion market uncertainties but exist to obscure the exorbitant profits. In their view, dealers and owners act in collusion to suppress the fish prices paid to boat fishermen. "There must be two tallies," complained Dalton. "One they prepare in front of our face, and the other they prepare for themselves." The retail commission levied on every kilogram of fish that changes hands between owners and dealers substantiates this assumption. This commission—whose size is held secret but amounts to Cr$30 ($0.06) per kilogram—functions effectively to divide owners and fishermen, and arouses the suspicion of bad intent.[9]

## *Technology and Economy of Scale*

The Camurim boat fishing mode is in a phase of stunted growth. There is enough money and knowledge available to exploit marine resources with new techniques but the relatively low fish prices impede the financial success of larger boats as the additional cost of new capital goods exceeds the increase in revenues.

The port of Camurim harbors thirty-two motorboats: five 10- to 11-meter boats, twenty-one 8- to 9-meter and five 6- to 7-meter boats. The two largest boats have 4-cylinder, 32 HP diesel engines, while the remaining thirty boats are equipped with 1-cylinder, 10 and 18 HP diesel engines. The five largest boats and seven medium-size boats are liners. These twelve liners have a deck which makes them most suitable for hook-and-line fishing, while the remaining twenty boats have open hulls which facilitate the casting and hauling of nets.

Liners offer more comfort to crews, have larger fish storage holds, and can thus stay for longer periods at sea than open hull multipurpose boats. Higher running costs oblige liners to make long fishing trips of seven to ten days to take advantage of their size and capacity. Multipurpose boats, on the other hand, can choose between handlining and netfishing, and thus adapt their catching strategies to the varying ecological circumstances. This flexibility is especially beneficial when the fish prices are low. Small and medium-size boats can considerably cut down on their operating costs by fishing with nets close to shore for one night without spending any money on ice, bait, or food.

The motorboats in Camurim have generally been bought in Pôrto Seguro, Ilhéus, and Salvador. There is enough timber at low cost available in Camurim but the harbor does not have a shipyard with the electrical machinery needed to construct boats. Major repairs such as replacing a deck, renewing ribs, or enlarging a boat are done in Camurim but these tasks take months of manual labor and could have been done in a few weeks with the proper equipment. Shipwrights work on the riverbank under improvised shelters of palmleaves with nothing more than axes, adzes, hammers, planes, chisels, and hand drills. There are three engine mechanics in Camurim but a lack of tools allows them to do little more than replace broken components with parts that have to be bought in Rio de Janeiro or Vitória.

Motorboats use the following fishing gear: handlines, set longlines, trawl nets, and gill nets.[10] Boat fishermen use the same handlines as canoe fishermen do. Some boats have set longlines which are identical to the set lines of canoe fishermen but which, instead of a dozen, have several hundred baited hooks. The set longline (*espinhel*) is especially appropriate for catching shark and is therefore not very popular among boat owners. Shark fetches a low price, although the fins may occasionally be sold for a good price to Japanese traders. Furthermore, the longline prevents crews from leaving for other fishing grounds because, without the use of nautical maps, they are unable to locate fishing gear in the open sea.

In recent years, many owners of small and medium-size boats have been purchasing trawl nets. A trawl net is a conical net with two wings that is towed through the water by a motorboat. The small trawls used in Camurim are 15 meters long, have a width of 13 meters at the opening and cost Cr$87,500 ($175). Boats in Camurim only use bottom trawls that drag across the sea floor to catch shrimp and prawns. This fishing method is lucrative for a few months at the beginning of the year and is mainly done by 6- to 9-meter boats.[11]

Netfishing is a very important source of income for small and medium-size boats. Boats use either stationary gill nets as canoe fishermen do, or they drift with the nets in the sea currents. Drift nets can cover larger areas of water than fixed nets and therefore may be more productive. Boats need at least thirty-five to fifty wide-mesh nets in order to give the crews a reasonable income.[12]

## Social Inequality at Sea and Ashore

Early in the morning when the tide began to rise, Geraldo's liner was loaded with ice. The truck backed up close to the riverbank and two muscular longshoremen carried each 40 kilograms of crushed ice in plastic cases down the slippery slope. A few children snatched some pieces of ice that fell off and licked them carefully. The crew lingered on the bulwark, still recovering from last night's binge.

"Embora, embora, embora" ("let's go"), shouted Isidoro, the captain. "Let's get out of here before the river mouth becomes too shallow. Last time we almost ran aground on the sand banks." Only the *gelador*, the fisherman responsible for preserving the catch, was busy loading the storage hold. He shoveled the ice into the corners and jumped on it with his black rubber boots to compress it even more.

The size of the hold determines the amount of ice and fish it can contain, and it indirectly limits the length of the fishing trip. Small boats carry 200 kilograms of crushed ice, medium-size boats take 500 to 800 kilograms, and liners have large holds with a capacity of 1,200 to 2,400 kilograms.

"Who selected the bait?" complained Isidoro. "These cases have more salt than sardines! How will we ever catch any fish when the bait falls right off my fishing hook! This is theft." He grabbed the 60 to 100 kilograms of sardines needed for a ten-day trip and lowered them in the storage hold. "Hey there, get the two oildrums and fill them with water." Drinking water was taken directly from the river, the same river in which further upstream a boat owner was changing an oil filter, two children took a swim with their dog, and some women were washing clothes.

After two to three hours, the boat was ready to leave. The calm trip to the estuary took only twenty minutes along a beautiful course with palm tree-laced banks and luscious mangroves. Our voices were drowned in the tac-tac-tac-tac monotony of the engine that echoed in the swamp. When we traversed the river mouth, the boat began to shake violently trying to get across the incoming waves. The sea was quite high and the white crests of breaking waves forewarned a rough trip. "You'd better go below deck, Dutchman," urged Isidoro, "because you're going to get wet." He looked at the compass and plotted our course.

Lying under the front deck, the beating of the boat on the water was frightening at the beginning. The deafening, low-key sound each time the boat battered the sea was impressive; my anxiety rose with every crack of the wooden ribs. From my narrow resting place I could see the dancing horizon and the flashes of green as we moved further off-

> shore. After about three hours of sailing, we arrived at the chosen fishing spot. I had dozed off a little, daydreaming with the heaving boat. Beneditto was sleeping. He had put his leg on my shoulder while Agenor lay curved on my right. The engine stopped and we crawled out. As I rose, I saw the lighthouse of Viçosa at a distance of maybe ten kilometers. The men took their positions on deck and prepared the fishing lines with bait.

Despite its small size, the boat has a social division of space that reflects the hierarchical relationships and chains of command among the crew (cf. Zulaika 1981:16–29, 46–51). Isidoro, the captain (*mestre*), is the only person to sleep in the motorcabin above deck. The fumes of burnt diesel oil penetrate the cabin and the noise of the engine may keep him awake, but his sleeping place is more comfortable than those of the fishermen are. The crew rest in narrow bunk beds below the forward deck. The torn mattresses are infested with fleas, cold water oozes from the storage hold into the cabin, and the air is heavy with the men's perspiration. The smell of fish seems to multiply in the cramped compartment. At night, cockroaches come out of hiding to nibble at the fishscales left under the fingernails of the sleeping men. Scarcity of fresh water contributes to the malodorous atmosphere and lack of hygiene aboard. We did not have a change of clothes during the entire period at sea and we barely washed our hands and faces in the morning.

The meals on the weeklong fishing trips are well prepared. Aside from fresh fish, the crew eats jerked beef, beans, rice, manioc flour, and crackers, with tomatoes, onions, peppers, red palm oil, and spices to add flavor. Poor boat fishermen confessed to me that they eat better at sea than at home. Most fishermen take turns preparing meals but crews on large vessels prefer to pay one member to cook. Meals on smaller boats are of course less elaborate than on liners, while on overnight trips fishermen take their own snacks. At mealtime, the entire crew assembles at the stern. The status differences that appear in the sleeping and working arrangements dissolve and the men share the same meal, and eat from equally rusty plates and utensils. On these occasions the crew become conscious of their presence in a total institution away from friends and relatives.

A boat on a fishing trip resembles a total institution with members of the same gender who conduct the principal aspects of life in the same place and in each other's presence, and who structure their activities as a function of the productive pursuit of their tasks (Aubert 1965; Goffman 1961:6).[13] Boat fishermen sleep, eat, and work for periods of up to ten days isolated from the community's social institutions.

The captain has the best place to fish. Isidoro fishes at the stern, even though the helm is near the bow. He has the best chance to catch fish because of the position of his fishing lines and baited hooks. Fishermen prefer sardines when fishing in the open sea because of its high content of oil. The oil disperses in the water and predators swim against the current toward the strong smell. The fish are more likely to encounter Isidoro's bait than that of any other fisherman. The least experienced or least productive fisherman remains at the bow while the others stay at the boards.

The social distribution of space also indicates the division of labor and the relationships of authority among the crew. Isidoro carries the responsibility for the safety of the crew, the condition of the vessel, and the success of the fishing trip. His decisions are intended to maximize the profit of the fishing enterprise and thereby indirectly benefit each fisherman individually. This objective gives him the authority to choose the fishing method, the location, and determine the duration of the trip.

Isidoro also decides about the composition of the crew and distributes all tasks necessary for the fishing operation. Small chores such as weighing the anchor, cleaning the deck, refilling the fuel tank, or setting the sail are done without complaint, but the more time-consuming tasks of storing fish, cooking, and standing guard at night are met with opposition, especially from ambitious fishermen who do not like to spend precious fishing time to render unremunerated services. A captain who does not want to lose his best men to other crews must seek some compromise. He will persuade a newcomer to prepare meals and preserve the catch for a small reward. Keeping vigil at night, however, has to be done by all. Liners fish often on the busy route of transatlantic cargo ships and many crews have experienced near misses because of watchmen who took a quick nap.

In some crews, the captain is also the navigator, while a helmsman (*motorista*) steers the boat and services the engine. Whenever possible, Isidoro tries to make difficult decisions through the consensus of the crew, but he is ultimately held accountable for the smooth functioning of the fishing trip and therefore has, at times, to endure the dissatisfaction of some crew members.

The general atmosphere aboard varies with the success of the fishing trip. During the voyage I accompanied, there was a lot of friendly joking when things went well. "I bet that you can't catch a larger fish than this one," said Batista as he pulled a 60-kilogram sea bass on deck. "Wait a second," yelled Agenor, "you won't have any place to put it because I'm going to top the storage hold singlehanded." But when no fish were caught, there was much backbiting and mutual irritation.[14] Isidoro told me a few weeks later about recurring incidents of outright sabotage. A fisherman who suddenly caught fish, during what seemed to be a bad night, would not call his sleeping comrades but instead tried to prevent the hooked fish from hitting the deck because it might attract the crew's attention. Some boat fishermen have cut the hooks of those who have dozed off, entangled their fishing lines, or have stolen fish.

Fishermen also fear more subtle but, in their eyes, infinitely more dangerous schemes to harm their performance. They are afraid of magic spells (*rezas*, literal meaning: prayers) and the damaging effects of the evil eye of envious fishermen (cf. DiStasi 1981; Maloney 1976). Some men prefer to stand near the bulwark, instead of sitting down, so that nobody can make a spell on their back. They are careful not to leave their sandals on deck when they take a nap, because someone might slip one sandal in the other to form a cross and place them upside down so that the owner will not wake up. As with all beliefs in complex societies, some people ignore them, but many fishermen and owners have their nets and fishing lines blessed to protect them from the evil eye. They also use antidotes that neutralize the bad effects of envy. Powders (*pemba branca, pemba preta*) with names such as Vence Tudo (Conquers Everything), Comigo Ninguém Pode (With Me Nobody Can), and Abre Caminho (Open the Way) are passed over the fishing lines. These remedies can be obtained from a spiritual leader of a Candomblé center. Ritual offerings to the Sea Goddess are another means to improve the fishing performance. All these spells, offerings, and sacrifices are based on a local notion of contagious magic called *simpatia* (sympathy).

Conflicts among crewmembers exist also on the small and medium-size boats but are less intense because fishing trips are shorter and the variation of fishing methods cushions tensions built up at handlining. Netfishing is a joint effort with equal shares for all fishermen. The long periods of rest between hauls and casts are spent playing dominoes and cards, or listening to the radio. The absence of leisure-time activities on liners prevents small quarrels from being resolved, and allows tensions to develop into major conflicts that can only be made up ashore with the generous exchange of drinks.

Beside displays of competitive strife on liners, there are outbursts of verbal aggression which are expressions of differences in work ethics and motivation among the crewmembers. The hierarchical inequality and the chain of command between captain and crew lead understandably to tensions, especially in a total institution comprised of men with a common occupation but with different perspectives and objectives. Friction becomes even more strenuous if the captain is also the boat owner. As a capital owner, he tries to pass on as many running costs as possible to the fishermen to enlarge his profit. His objectives are to redeem his loan and to remain in the dealer's favor with good catches.

Conflicts between captain-owners and crewmembers appear in the form of insults, teasing, and derogatory remarks. Duarte (1981) describes similar forms of verbal aggression among purse seine fishermen in Rio de Janeiro as a pecking order among ranked men. He demonstrates how canoe fishing crews, with their spirit of reciprocity, equality, community, and comradeship have changed into purse seine crews with hierarchical relationships, unequal remuneration, and rigid class distinctions. The hostility among the crew, he argues, expresses the contradiction between past values of cooperation and current inequality and atomism.

Insults among boat fishermen in Camurim are not structured along the rungs of a pecking order, possibly because handlining is not a cooperative activity and the individual incomes depend more upon personal success than upon one's status in the production unit. Tensions arise, however, from an informal division between high-producing and low-producing fishermen. Twenty of the one hundred and twenty boat fishermen are considered high-producing, industrious, and ambitious by the boat owners and their colleagues. They are *bom de linha* (good with a fishing line). Most fishermen state that the high-producing fishermen are more skillful and active than others, while many low-producing fishermen are simply poorly motivated or have physiological inhibitions at sea, such as sleepiness, tiredness, lack of appetite, and even seasickness. The best fishermen attribute part of their good fortune to insomnia. Luck is acknowledged as a factor in success but the men believe that good catches will not last without hard work and superior skills (cf. Kottak 1983:90–95).[15]

Forty fishermen are regarded as low-producing. The remaining sixty men occupy an ambiguous place in the social hierarchy. These men consider themselves as hardworking and ambitious but owners and high-producing fishermen classify them among the low-producing group. They barely break even on most fishing trips and sometimes get into debt. Unless indicated otherwise, I will include this group among the low-producing fishermen because their chances of becoming boat owners are negligible, and much of their behavior ashore and at sea resembles that of the forty permanently indebted fishermen.

High-producing fishermen are aggressive in maximizing their performance at sea, and in disputing imposed tasks that do not contribute to their income. Work is seen as an opportunity for economic mobility. The fisherman enters into a voluntary and periodically renegotiable verbal labor contract that does not have stipulations about the size of the revenues. This arrangement may give the high-producing fisherman the chance to save enough money to eventually obtain a loan to buy a boat. Therefore, he wants to take as much advantage as possible of the fishing trip, remain at sea when fish are being caught, and quickly leave for other fishing grounds if unsuccessful. The disputes with the owner about the catch division and the submission demanded are ways to renegotiate his labor contract and improve his possibilities for economic ascendence.

For the low-producing fisherman, work is an obligation, a necessary corollary of his submission to the boat owner. He is in perpetual debt and therefore does not have the incentive to work hard. He clashes with high-producing fishermen on decisions that prolong his stay at sea and raise the operating costs. The debt relationship keeps him permanently tied to the boat owner and obliges him to fish. Only work, regardless of its yield, allows him to demand money for household expenses. Little can stop him from leaving the crew and not paying his debts, but such action will damage his already poor reputation. Boat owners will be reluctant to employ him, and his credit in local stores will drop even further.

Boat fishermen seldom have enough savings to pay for the household expenses during a long fishing trip. Hence, the boat owner pays each fisherman an advance (*vale*) on his yet-to-earn income before embarking. Credit is extended if the boat cannot leave because of bad weather, repairs, or insufficient supplies. Most open confrontations ashore between fishermen and owners revolve either around the credit advance system or the catch division. The boat owners are concerned about profit, the indebted fishermen about the credit advance, and the high-producing fishermen about the share system.

### *The Division of the Catch*

Around midnight on our sixth day at sea, Isidoro said that it was time to go home. He was disappointed about his poor performance but the other five crew members had caught enough to make the trip worthwhile. "The tide will start to rise at four in the morning, so we should be leaving soon."

The fishermen rolled up their fishing lines, closed the storage hold and removed all loose objects from the deck. Isidoro cranked the engine, and the crew retired to their bunk beds.

At 7:30, I could clearly see the river mouth frothing on the coastline as we approached Camurim. I had my sweater and coat on, although the sun was already quite high. The night had been cold and the fresh morning wind kept me from getting warm. Isidoro blinked his eyes as the sea water splashed in his face. "Everybody to the bow because the mouth is really shallow today." We counterbalanced the engine with the weight of our bodies and were drenched by the waves as we surfed past the sand banks. Once we entered the river, a hot humidity descended on us that made us take off our sweaters and long for a cold glass of beer.

Geraldo, the boat owner, had seen the mast of his boat above the mangroves and he was already waiting on the riverbank. "You're early, the truck isn't here yet. But let's go ahead and unload the catch." The longshoremen carried the catch ashore and emptied the boxes on the cement floor as the crewmembers separated the fish into piles according to distinct marks on the throat, a gill, or the upper, lower, or both tailfins of the fish. The captain leaves his fish intact.

All fishermen showed up to witness the weighing after the catch had been sorted out. Geraldo and the fish dealer stood behind the scale and wrote down the weight and species of each individual catch. A small crowd of old men and poor children gazed over the shoulders of the crew waiting to receive some small fish. After the catch had been loaded into the truck, the dealer, the boat owner, and the crew left for a nearby bar to settle accounts.

Boat fishermen are entitled to take fish for their own consumption at home. The amount has been established at 3 to 5 kilograms because owners suspect fishermen of taking more fish than they can eat and selling the remainder in town. If a fisherman wants some extra fish, he has to buy it from the owner. Fishermen succumbed to this restriction only after much protest because they feel that they are inalienable from their catch. They consider themselves self-employed producers, while boat owners consider them laborers. The owner states that he pays half of the running costs and thus has the right to one-half of the entire catch. He buys the catch from the crew and resells it to the dealer, earning a commission in the transaction.

The separate line share system is used for all handlining catches. There are numerous variations because of continuous contention between owners and fishermen about the inclusion of certain costs or about the size of the shares. Captains, navigators, helmsmen, or mechanically inclined fishermen argue that their unique abilities should be rewarded with larger shares of the catch. The basic principles of the catch division, however, are never fundamentally changed. Following a set of rigid rules relieves boat owners of a social responsibility for their employees.

The division of the catch made during the trip in November of 1982 illustrates the share system. First, the operating costs are determined (table 4.1). Other recurring items that do not appear on this list are motor oil, oil filters, brooms, baskets, pots, utensils, fishing weights, and butane gas for the stove. Owners say that the fishermen will be more careful with the equipment if the costs are shared among them. Fishermen complain that they have to pay for goods that pertain to the standard equipment of the boat, while they are not entitled to a share of the fishing gear when they leave for another crew. Therefore, fishermen sometimes take fishing lines, food, and butane containers home, claiming that they are rightfully theirs.

The running costs were equally divided among the fishermen and deducted from the gross value of their individual catches (table 4.2). The net value was divided in two: one part for Geraldo as his return-on-capital and one part for the fishermen's labor. There were some extra costs and revenues. Beneditto received Cr$300 ($0.60) from each colleague for preparing food. Lorival earned Cr$2 per kilogram from the dealer for preserving the catch: 519 × 2 = Cr$1038. Isidoro received some extra shares as the boat's captain, navigator, and helmsman. These shares serve to induce captains to work hard. The extra revenues were deducted from Geraldo's part of the catch. Several owners give their captains a larger share when the total catch is low. It is their

*Table 4.1*

Operating Costs of an Eleven-Meter Motorboat After a Ten-Day Fishing Trip by a Six-Man Crew

| | |
|---|---|
| ice (2,000 kilograms) | Cr$ 10,000 |
| bait (100 kilograms) | 15,000 |
| diesel oil (275 liters) | 22,000 |
| anchor | 2,000 |
| fishing line | 2,240 |
| hooks | 1,750 |
| food | 13,095 |
| medicine | 350 |
| TOTAL | Cr$ 66,435 ($133) |

acknowledgment that not a lack of skill but unforeseen natural circumstances caused the poor catch, and is thus a reaffirmation of the owner's trust in the captain's ability. In this case, Isidoro received $2\frac{1}{2}$ extra shares or $2\frac{1}{2} \times \frac{1}{12}$ of the total net value (had all men caught the same value in fish, then each fisherman would have earned $\frac{1}{12}$; the captain $3\frac{1}{2}$ times $\frac{1}{12}$; and the owner also $3\frac{1}{2}$ times $\frac{1}{12}$ of the catch). During this trip, Isidoro earned $55{,}920 \div 12 = 4{,}660 \times \frac{1}{2} =$ Cr$11,650 extra (see table 4.3).

The boat owner earned $27{,}960 - 11{,}650 = 16{,}310$ plus Cr$30 a kilogram as retail commission. His profit was $16{,}310 + 15{,}570 =$ Cr$31,880 ($64). The fishermen did not receive the income

*Table 4.2*

Net Value of the Individual Catches of Six Boat Fishermen After a Ten-Day Fishing Trip

| | *Catch (kgs)* | *Gross Value* | | *Expenses* | | *Net Value (cr$)* |
|---|---|---|---|---|---|---|
| Isidoro | 35 | 5,800 | – | 11,072.50 | = | –5,272.50 |
| Agenor | 158 | 38,700 | – | 11,072.50 | = | 27,627.50 |
| Batista | 96 | 26,150 | – | 11,072.50 | = | 15,077.50 |
| Lorival | 95 | 19,430 | – | 11,072.50 | = | 8,357.50 |
| Beneditto | 79 | 16,350 | – | 11,072.50 | = | 5,277.50 |
| Rúbi | 56 | 15,925 | – | 11,072.50 | = | 4,852.50 |
| TOTAL | 519 | 122,355 ($245) | | 66,435.00 ($133) | | 55,920.00 ($112) |

**Table 4.3**

Income Division Among Six Boat Fishermen After a Ten-Day Fishing Trip

| | *Net Value* | *Balance* | *Cook* | *Gelador* | *Navigator* | *Income* |
|---|---|---|---|---|---|---|
| Isidoro | −5,272.50 | −2,636.25 | −300 | | +11,650 | = 8,713.75 ($17) |
| Agenor | 27,627.50 | 13,813.75 | −300 | | | = 13,513.75 ($27) |
| Batista | 15,077.50 | 7,538.75 | −300 | | | = 7,238.75 ($15) |
| Lorival | 8,357.50 | 4,178.75 | −300 | +1,038 | | = 4,916.75 ($10) |
| Beneditto | 5,277.50 | 2,638.75 | +1,500 | | | = 4,138.75 ($8) |
| Rúbi | 4,852.50 | 2,426.25 | −300 | | | = 2,126.25 ($4) |
| TOTAL | 55,920.00 | 27,960.00 | | | | 40,648.00 ($81) |

noted above because the credit advance given before the trip had to be deducted. All fishermen received Cr$4,000 ($8) and Isidoro Cr$6,000 ($12) before embarking, leaving Rúbi in debt and Beneditto with very little money. This particular example shows a good catch during a bad summer fishing season. Most catches are worse and inevitably lead less successful fishermen into heavy debt.

Netfishing and longlining are cooperative efforts. Catches are therefore divided in equal parts among the fishermen. After deducting the running costs and an additional 3 to 4 percent for net repairs, the owner of the boat and net gets half or one-third of the net value and the fishermen share the remaining half or two-thirds. The owner, like any other fisherman, receives a share if he participates in the fishing effort. Captains on small and medium-size boats hardly ever receive extra shares. The return-on-capital for most of these boats is one-third (*terço*), whether the crew went handlining or netfishing. Boat owners are trying to take half of the catch as their share but the resistance among the crews is great.

Leaning on the bulwark of the small boat as the mechanic tried to repair the engine, a discussion developed between the owner, Bernardinho, and the two fishermen, Tonho and Augusto. Bernardinho was irritated because the fishermen had ruined the engine by pouring cold water on the overheated radiator.

> "This business of the terço can't go on any longer," he began. All large boats use a *meia-meia* [50-50 division] but the group of the smaller, open-deck boats are addicted to the terço." More vehemently than usual, Tonho answered back, "a boat owner should also fish so that he can feel the struggle at sea." "That's right," echoed Augusto, "and a captain-owner should use his share as a fisherman to live on and save the terço to pay off his debts and save money for repairs. He shouldn't try to live as a *dono* [absentee owner] and pretend that he is rich." Bernardinho realized that these remarks were directed at him and he replied angrily, "You two are going to pay for this, I will deduct the costs of this repair from your shares."

The two fishermen had been pointing out that the catch division is profitable as long as the owners of small boats participate in the production effort and do not act as absentee owners with expensive life-styles. In their eyes, Bernardinho does not have the status of the owners of liners who command a much larger investment and obtain higher profits which allow them to claim the prestige of the landed elite.

### *Boats and Dependence*

It was one of those lazy, dusty Saturday afternoons in late-summer during which a group of fishermen and I were resting in one of the many small bars in the harbor quarter. An oriental-looking man stuck his head in the doorway. He mumbled some greetings and asked if anybody had shark fins for sale. "No, we don't, but maybe Braga has some left, my friend." The man said thanks and continued on his way through the deserted streets without asking where this Braga lived. "Antônio," asked Faustino, "what do these Japanese do with those shark fins?" "I don't know," I replied evasively, "what do you think?" "Well," he said, "they say they make soup with it but I don't believe them. Why would they pay four times as much for shark fins as for snooks? They could make a more delicious soup from sea bass or sea trout." "You know what I think?" interrupted Isidoro. "I think that there must be some drug inside these fins. Why would they otherwise buy those smelly fins for such a high price?" "I think that they make dynamite from it," said another, "or it must have something to do with nuclear energy." Just as he said this, Everaldo the mechanic walked in. "Ah, you're talking about the shark fins again. I'll tell you what the Japanese use them for," said Everaldo with confidence. "You know that when you're out at sea and you look on the *sonda* [probe] that you can see a school of sharks, right? Well, that means that they are radioactive, because the sonda is like a radio. Now, the Japanese buy these shark fins to extract their radioactivity and with this they make atom bombs with which they want to conquer Brazil!" He looked triumphantly at all of us and asked for a beer.

During the following days I solicited some more explanations about the enigmatic shark fins. Braga, who used to buy fins from the fishermen for resale, was very matter of fact about it. "Shark fins are used for nuclear energy because the shark has a high speed. He is very fast." Salazár, the public accountant who had had a high school education and was considered as someone who knew about the things of the world, laughed heartily about all these explanations. "No, not at all. Nuclear energy, dynamite, drugs . . . ," he chuckled. "When the Japanese buy these dried and lightly salted fins, they submerge them in water and extract some thin white fibers from them. They use these fibers for computers. I once saw the inside of a computer in Salvador and it contained lots of these white threads. If they were to use it for something illegal, then these Japanese traders would be in jail right now."

The enigma of the high-priced shark fins is indicative of the mysteriousness and unintelligibility of the urban market for boat owners and boat fishermen. Just as they believe that the Japanese are secretive about the real use of shark fins, that they are aware of their high economic value, so they believe that urban dealers and bankers mystify the workings of the economy.[16] The interpretations of boat owners and boat fishermen about the interactional and institutional operation of the elusive economy is therefore aimed at enhancing their understanding. The boat figures prominently in this discourse. Boat owners, high-producing fishermen, and low-producing fishermen each in their own way believe that the boat creates dependence and submission.[17] The boat is not symbolic of their social position, but its operation and the infrastructural support needed to produce and market fish actually entangle them in hierarchical relations. The boat is a cultural focus which shapes its economic context and the society around it through its reproduction.

As has been described in detail in this chapter, boat owners perceive a structural dependence on banks, loan funds, communication systems, suppliers, urban markets, and other national institutions. Dependence is also a structural characteristic of boat ownership through the functional operation of the production unit. Boat owners may be more powerful than boat fishermen but can only produce with fishermen's labor. The owners feel wronged by the fishermen—who accuse them of exploitation—and by the fish dealers—who fail to show compassion for their financial difficulties. "The fish dealer works for the market," complained Tiago, the boat owner, "but we are offered a fixed price. He gives us ice free of charge so that he can receive the fish. The boat already leaves the port in debt, and in this way we become the employees of the dealers."

The need for the steady distribution of fish to the urban markets has influenced boat owners' personal relations with wholesalers. Deference and sometimes even submission are the interactional expressions of structural dependence. Boat owners have surrendered themselves to impersonal economic forces, whose mediation through dealers they try to influence in face-to-face encounters. They feel obliged to treat the dealers with a deference that is more basic than the respect with which they approach their financial guarantors. Boat owners believe that their status difference with the elite will disappear in time but that the dependence on market relations will persist for many years to come.

The attention of boat owners is directed to the consolidation of their enterprises and the legitimization of the claim to elite status. Capital accumulation is a step toward higher status in all walks of life. Boat owners want to be part of the social and political elite of Camurim and have the prestige and esteem bestowed upon them which they believe they deserve as capital owners. Just as readily as they complain about their dependence on dealers, several retort with confidence that "I am independent. I do not owe anybody anything."

High-producing fishermen interpret their dependence within the economy as interpersonal. They are not concerned with economic institutions but have an everyday experience of the need for capital and work, of the insecurity of employment and income, of the uncertainty of catches and prices, and of the conflict between their own ambition to advance on the economic ladder and the boat owners' persistence in protecting their own socio-economic position. Economic mobility cannot be achieved just by personal determination or through appeals to banks and federal agencies, but relies on the assistance of members of the local elite.

Dependence is manifested from day to day through a subjugated position in economic practice. Boat fishermen do not have the submissive, almost dejected, attitude that is so characteristic of plantation workers in the presence of a landowner, but they have to submit to the orders and demands of boat owners and captains to retain their position in the fishing industry. Boat owners consider deference as an interpersonal strategy to cope with the structural dependence of the fishing unit in the national economy. High-producing fishermen have been socialized into subjugation and regard the alleged structural dependence as a negotiable condition. They argue that boat owners are independent, in control, and on equal terms with the dealers.

For high-producing fishermen, the boat offers the possibility of economic ascendence. They focus on distribution. Only a fair share of the catch will give them the chance to acquaint themselves with the elite during informal encounters in bars and clubs. Their goals lie in the economic organization, not in the spheres of politics and public acclaim. They desire the patronage that boat owners are trying to overcome.

Low-producing fishermen have a third perspective on the boat as a formative force in their lives. They acknowledge most of the implications mentioned by high-producing fishermen but point less to inhibitions to economic mobility and more to the social consequences of economic practice. The credit advance system, the indefinite indebtedness, and especially the absorption of the household into the production unit describe better what low-income fishermen mean by dependence. The survival of the household depends on the boat owner's willingness to extend credit. The low-producing fisherman plays a marginal role in the economic maintenance of his family, irrespective of the temporary success of his productive efforts. The boat owner has taken his place as the provider of the household, as will be explained in chapter 7. Submission is more a state of social inferiority than a strategy or structural characteristic. Low-producing fishermen undergo their subjugation to owners, captains, and high-producing fishermen with an awareness of their position at the bottom of the social hierarchy. They think of high-producing fishermen as upwardly mobile and boat owners as part of the elite, and contest claims of dependence by either group as hypocrisy.

Low-producing fishermen are disillusioned with the boat fishing industry, which signifies indebtedness for them. Yet, the boat is the only means of production that has the potential to erase all debts with a few lucky catches, and such good fortune may reap a switch to canoe fishing. They waver between the canoe fishermen's desire for economic stability and security, and the boat fishermen's willingness to take greater risks to achieve their social and economic aspirations.[18]

# 5

# Boats and Canoes in Coexistence

Leonardo warned Jorge when he passed his house with a small bundle of clothes in one hand and his knife in the other. "Fishing for others doesn't pay, Jorge!" he shouted. Jorge shrugged his shoulders and continued on his way to the harbor. I had talked to Jorge the day before about his decision to go boat fishing. He told me that he had problems finding a reliable fishing partner and that the catches with his canoe had been disappointing. Winter was the good season for fishing far offshore, so he had accepted an offer from a boat owner to join his crew for some time. When Jorge was out of sight, I asked Leonardo about his admonition, "Why did Jorge switch to the boat?" "Because he has no sense of responsibility," replied Leonardo. "Casimiro already warned him the other day that he should take care of his nets because they give him his independence, but he doesn't listen. Now, he is subjecting himself on a boat" (se está sujeitando num barco).

Canoe fishermen and boat fishermen hold strong opinions on why their own fishing mode is superior, yet some fishermen like to switch back and forth between boats and canoes. Canoe fishing and boat fishing may constitute separate modes of production with proper fishing grounds, production units, technologies, distribution channels, and price setting conventions, but they are bridged by an interchange of labor. The fishermen consider the production modes as part of a pluriform fishery, and their active involvement in one is supplemented by a strong opinion of the other. Their vision of the fishing economy of Camurim incorporates the other production mode as a contrasting, and possibly alternative, form of employment. Boat owners and high-producing fishermen hold a negative opinion of canoe fishing, while a group of low-producing fishermen have a positive perception. Canoe owners reject boat fishing outright, while many canoe fishermen view it as a viable alternative in winter.[1]

The switching between production modes gives canoe fishermen and boat fishermen the opportunity to compare the two modes and convey recent changes to their colleagues. Switching is not simply the exchange of one economic task for another. A fisherman finds himself in a new context which may change his interpretation of the production modes, reformulate his objectives in life, and obscure the initial intention for switching. This process of change leads to a tension between practice and identity. People cannot immerse themselves in the practice of one fishing mode and maintain the social identity associated with another without serious inner conflicts.

Although I stated in chapter 3 that social identity and economic practice are closely related, this does not mean that a new economic position will immediately effect a change of identity. Canoe fishermen underwrite the low-producing fishermen's interpretation of the dependence and submission of boat fishing. However, they also believe that when properly taken advantage of, boats have a greater earning potential than canoes. Only persons who work permanently on boats will become indebted. These canoe fishermen want to raise their income quickly during the good boat fishing season and then resume canoe fishing when coastal resources replenish.

Canoe fishermen who switch to boat fishing have to conform to the social interaction prevalent among the crews. The production mode obliges them to become what they do not want to be, and to act in ways that are in conflict with how they see themselves. This discrepancy between practice and personal identity is justified as a sacrifice in benefit of a more important end. Submission and dependence are temporary concessions to earn enough money to sustain their households at a desired standard of living. Yet, after several fishing trips, it becomes increasingly difficult to sustain this motivation and retain their former personal identity. Canoe fishermen cannot simply confine the implications of the boat to the economic domain and continue to be free and independent in other walks of life. The objectives that motivated the switch from canoe fishing to boat fishing recede to the background, while new demands become relevant. The boat and the bar replace the canoe and the house as their cultural foci.

After a period of time, production activities and consumption patterns contradict the fisherman's own image of independence and autonomy. His former canoe fishing colleagues are among the first to detect this change. They feel that he no longer shares certain fundamental social attributes with them, and they begin to identify him with the boat fishermen (cf. Erikson 1959:102). A discrepancy has arisen between a self-image grounded in canoe fishing and a social identity to which the switching fisherman tries to hold onto discursively. This problem can be ignored for some time but has to be resolved eventually, because increasingly he begins to see himself as a mirror image of the men who surround him (cf. Luckmann 1983). This breakdown of the self, which is similar to a breakdown of routine behavior because both are reproduced in practice and resumed after a discursive conflict, will eventually lead to self-reflection and make the fisherman choose between canoe fishing and boat fishing.[2]

This chapter describes first the social and economic interaction between canoe fishermen and boat fishermen, and how the places where they reside, the separate harbors, and the location of public services facilitate their segregation. This social separation of canoe fishermen and boat fishermen and its effect on the identity formation in childhood are important factors in perpetuating the two modes of production.[3] Stereotypes and feelings of superiority reinforce the social boundaries of the two groups. A description of the different interpretations of the pluriform economy and an analysis of the preference of one fishing mode over another clarify how these perceptions are in conflict. An income analysis demonstrates that catch statistics, which all acknowledge as correct, lose their objective status when they are interpreted by fishermen who use different criteria to determine the differences. A brief history of the emergence of the pluriform economy shows why canoe fishing did not disappear and why it continues to be a viable alternative to boat fishing. An analysis of uncertainty as the overarching meaning of fishing shows that each social group proposes a different strategy to deal with fluctuating catches.

## *Social Interaction Among Fishermen*

> When Isidoro and Casimiro met on a Saturday morning in mid-June of 1983 to discuss the sale of a canoe, it seemed as if it was a rendezvous between old friends who had gone separate ways and for many years had lived in different towns. They were courteous to one another and avoided the usual boisterous talk laced with coarse words that is so common among fishermen. "Casimiro, do you remember when we would run away from school during the morning break and go to the woods to hunt small birds with a catapult?" "Of course, and then we would get thirsty and climb across the fence of old João's palm grove and steal coconuts." "How he beat us when he caught us one day!" "Sure," interrupted Isidoro, "those were good times . . . there was more fish then also. The beach seines would get so full that we would throw all small fish to the poor."

Boat fishermen and canoe fishermen know each other by name from their schooldays and encounters at sea, but ashore they see each other seldom. They live in different neighborhoods divided by the main street of Camurim. They do not have to enter the other side because the principal public services are located along the central thoroughfare. Furthermore, each neighborhood has grocery stores, bakeries, butchers, and bars to satisfy all daily consumption needs. Even the purchase of bait does not require any direct contact because canoe fishermen send their fish peddlers to buy shrimp. "Outside the obligation to help them when they are dying at sea," said Casimiro after he had sold his canoe to Isidoro, "I stay here and leave them over there alone."

The socio-spatial boundary set by the main street is reinforced by the absence of affinal and consanguineal relationships. The two neighborhoods, which are of course not composed only of fishermen and their families, are almost endogamous. Endogamy is not a prescribed marriage requirement, but the lack of interaction between the neighborhoods makes men choose future spouses from their immediate vicinity.

The poor interpersonal contact and social integration of canoe fishermen and boat fishermen are not fully extended into the economy. Economic goods, technical skills, knowledge, and labor circulate quite freely among canoe and boat fishing, and developments in one production mode will have effects on the other. For instance, the ratio between medium-size and large boats has long-term consequences on the revenue of canoe fishing because the composition of the fleet determines the degree of exploitation of the shallow coastal waters. Also, although fish prices are established differently in the two production modes, a sharp price fluctuation in one market will have an effect on the distribution of fish produced by the other mode. High urban demand for fish around Easter prompts canoe fishermen at times to sell fish to the wholesalers, and some fish caught by boat fishermen find their way to the tourists in summer. The most important connection between the two fishing modes is labor. Fishermen can work both on boats and on canoes because the fishing techniques vary little.[4] Aside from situations beyond the fishermen's control, such as the depletion of tidal zone resources that obliterate beach seining, increased lumber demands that raise canoe prices, or inflation and a trade imbalance that affect fuel costs which might force fishermen into another fishing mode, the choice between boat fishing and canoe fishing is taken by the fishermen in terms of their perception of the pluriform fishing economy and how it might satisfy their needs and aspirations.

## *Mutual Perceptions Among Fishermen*

Canoe fishermen and boat fishermen refer to one another as turmas, the same term used for corporate groups. Canoe fishermen are called the turma do norte (the north), turma da praia (the beach), turma do batelão (the canoe) or turma da água suja (the dirty water). Boat fishermen are identified as the turma do sul (the south), turma do rio (the river), turma do pôrto (the harbor), turma da lancha (the boat), or turma da água limpa (the clean water). This division is more than a simple occupational classification because each term indicates a different context. The terms "northern" and "southern" turmas refer more to residence than to profession, because canoe fishermen in the south are also included among the turma do sul and northern boat fishermen pertain to the turma do norte. "Beach" and the "harbor" serve to describe the principal gathering place before and after work. "Canoe" and "boat" point to the fishing techniques, the work rhythm, and the owner-fisherman relationship. The terms "clean" and "dirty" water indicate the ecological zones and their potential revenues. Dirty water refers to the sediment-laden coastal fishing zone; clean water, to the high seas. The descriptive terms used in speech vary with the topic of discussion, as in the following conversation. "The men of the turma da lancha, Antônio, are always in debt. They are the owners' slaves." "You know what the problem is, Antônio?" said the other, "the turma da água limpa just doesn't catch enough fish near the continental slope."

Boat owners and boat fishermen who live near the harbor have a quite stereotypical picture of their northern colleagues. They rely principally on their own canoe fishing experiences from an already distant past. Canoe fishermen are considered to be "primitive, backward, and provincial" (*primitivo, atrasado e rústico*). Instead of embracing the technical advances of motorization, trawl nets, and new fish preservation methods, the northern fishermen prefer to use their "physical strength against the forces of nature." "I prefer to see the engine wear itself out, not myself," said Júlio about the backbreaking work of canoe fishing. The boat fishermen see themselves as the true heroes of the ocean who will spend days on small boats on the high seas. Many have experienced the danger of storms, but they have also felt the euphoria of catching hammerhead sharks weighing 100 kilograms and earning large sums of money in a single day.

Canoe fishermen are better informed about the lives of boat fishermen because many continue to fish on boats. "The boat fishermen talk very rudely," said Natalício. "They're always swearing and hollering as if they were having a fight. It seems as if they belong to a different nation." Having spent a lot of time with boat fishermen in bars, I tacitly agreed with him. Yet, the boat fishermen themselves were aware of this perception. After an evening of heated exchanges among a group of boat owners and boat fishermen, one boat owner took me aside and said, "It may seem that we are always fighting but that's only an impression. Arguing is a way to confirm our friendship because soon after we forget our disagreements."

Canoe fishermen seem always a bit intimidated by the boisterous demeanor of the boat fishermen. They regard boat fishermen as "drunkards, unloving husbands and fathers, ignorant, irresponsible, and captives of the boat owners. They like their friends better than their own family." Canoe fishermen emphasize their own dedication to their wives and children, and their independence and self-reliance as producers.

Both groups of fishermen explain the differences in terms of temperament, personality, and personal choice. The most common argument is that people are what they are because they choose to be that way or because it lies in their nature.

The general assembly of the Fishermen's Guild is the only formal occasion at which all fishermen and owners can meet one another. The Guild is dominated by a president who, with the support of some boat owners, has not held annual elections or organized a reunion for several years.[5] Most reunions held at the Guild are recalled with resentment by the canoe fishermen. The Guild is clearly controlled by the southern fishermen who sneer at any dissenting policies proposed by the northern colleagues. Canoe fishermen also complain about the location of the Guild near the harbor. They desire a more neutral place along the main street because the office has become an informal club for some boat owners and boat fishermen. They prepare meals, play cards, drink, and even store fishing gear at the site. The Fishermen's Guild has also become a retail outlet for low-quality fish which wholesalers do not want to transport to the city.

> The Guild became involved in the sale of fish in 1976 when, according to Eliseu, the prefect forbade fish peddling in the streets of Camurim. "He was trying to attract tourists to Camurim and thought that fish hawking gave the town a dirty and uncivilized appearance. All fish had to be sold at the Guild." Macário, a canoe owner, gave another version. "This affair was a collusion between Eliseu and a corrupt politician attempting to control the trade of fish. Canoe fishermen who brought fish to the Guild were paid with worthless *vales* (IOUs). When I took two snooks to the Colônia, Eliseu smiled at me and said: 'That's just what I was waiting for.' He took the fish from my hand, sold them to a tourist, and gave me a *vale*. 'What the hell is this?' I said, 'Why don't you give me the money?' 'Oh,' he said, 'I need the money to go to the brothel tonight.' I called him a dirty thief, took my knife from my belt, and grabbed the money that was in the cash register. I give him my *vales*, and vowed never to return again. The next day I made a good catch and sent my own son in the street to sell it. An hour later, me and my fishing companion were arrested. The bastards. We were thrown in jail with two murderers who had killed a cocoa planter near the waterfall upstream. Luckily, the same day a lawyer succeeded in releasing us from jail and the ordinance was withdrawn."

This conflict and the one-sided reunions have enhanced the mistrust between canoe fishermen and boat fishermen, and have reinforced the social boundaries between the two production modes. The only occasion at which I felt a strong feeling of unity and great compassion among all fishermen was when a boat had been lost at sea.

> It had been drizzling steadily for a couple of days but most captains thought it safe to go fishing. On Thursday morning the weather turned foul. A cold southern wind from the Antarctic swept across Brazil and dropped its storm on Camurim. All boats were safely home before dusk, except Everaldo's 11-meter liner. Strong gusts of wind battered the wooden shutters of the houses the next morning, and there

were frequent power outages. Eliseu dispatched messages to other harbors to see if Everaldo had taken shelter there. The owners of the largest liners organized a search party when the storm subsided on Friday. Within hours they found the vessel. The boat's engine had broken, the sail had been torn, and they had lost two anchors. The crew had spent two days below deck while the boat drifted aimlessly at sea. The riverbanks were crowded with people cheering when the lost sailors arrived, and candles were lit in Church for their safe return home.

### *The Choice and Preference of a Fishing Mode*

When I asked what they considered the two most important things in life, canoe owners and canoe fishermen mentioned health and family most frequently, and boat owners and boat fishermen named money and friends. These statements should not have too much importance attributed to them, but they do represent a gut reaction that can serve as a first approximation to a better understanding of social and economic expectations and what motivates these men to choose one fishing mode over another.

One canoe fisherman expressed the different expectations between canoe fishermen and boat fishermen as follows, "Who desires doesn't sleep, who doesn't desire sleeps well" (Quem quer não durme, quem não quer durme bem). In other words, those who have limited demands live in affluence and tranquility, satisfied with what they have. But those who are ambitious and set their goals very high are always preoccupied, worrying about debts and installments. The two groups seek to satisfy different demands and will interpret the economic practices with those objectives in mind. Scarcity and affluence, as Sahlins (1979) and Bataille (1988) have argued, are not objective but interpretational evaluations of economy and society.

Canoe fishermen say that sleeping and eating at home is the principal reason for their preference of shallow water fishing. "Years ago I went fishing for a couple of months on a boat," explained Natalício, "and I'll tell you, it wasn't right for me. The nights of sleep lost, the irregular meals, you don't know when you will have something to eat, the absence of a woman, and the longing for my family made it hard to stay at sea for a week." The canoe fishermen strongly believe that infrequent sexual relations will deteriorate one's physical strength and manhood, and that poor hygiene and lack of sleep harm one's body. The physical and spiritual health of the boat fishermen and the well-being of their families suffer disproportionately to the supposedly higher earnings, according to the canoe fishermen. Casimiro told me of the time that his wife was pregnant and he could not stop worrying. Other canoe fishermen told of similar experiences when they had to leave a sick child behind as they went on a long fishing trip without being able to offer any assistance if the child's condition worsened. They stated that boat fishermen place the household's material prosperity above its emotional well-being. The father has less influence on the education of the children and his honor is constantly in danger because of the greater opportunity for his wife to be disloyal.

As is common in Latin cultures, men are convinced that women can be seduced easily and therefore should be confined to the house and avoid contact with unrelated adult males. The periodic absence of boat fishermen makes adultery more likely than among the ever-present canoe fishermen and imbues the conjugal relationship with mistrust between spouses.[6] Being away at sea prevents any control, and the wife is obliged to develop a network of men to help her in case of an emergency. Boat owners, wholesalers, storekeepers, and politicians are the most obvious contacts sought after by women whose husbands are at sea. The differences in status, gender, and wealth leave much room for improper proposals.

There are also economic advantages to canoe fishing over boat fishing. Canoe fishermen claim to be less susceptible to the weather, to be more independent, and to earn more money. Boats may be stuck in the harbor for days before the weather improves. Canoes can cast nets less than 50 meters offshore so that an hour a day of calm winds allows the men to fish. During my stay in Camurim, I counted only four days on which all canoes were beached, as opposed to an accumulated total of three to four weeks that prevented boats from leaving port.

The most important reason that canoe fishermen prefer canoes to boats is the independence of the owners and fishing partners. Canoe owners and canoe fishermen place much value on their independence. They interpret the credit advance system as a deliberate attempt to entice fishermen with easy cash in order to bind them permanently to the crew through never-ending debt relationships. Gilson summarized the relation between boat owners and fishermen aptly, "The boat owners give the boat fishermen the conditions to live because they keep giving them credit. The boat fishermen will never die of hunger because the boat owner will always give them money to buy food. This turns them into slaves because they are forever wrapped in a debt relationship with the owners."

Finally, canoe fishermen are convinced that their annual revenues are higher than those of boat fishermen. The living standard of the household is taken as their only measure, without ever taking into consideration the higher drinking expenses of the boat fishermen.

Boat fishermen have an entirely different outlook. Many acknowledge that they live through rough moments at sea but that they find much emotional gratification from the camaraderie aboard ship. Away at sea, far from the preoccupations of the household and pending debts, they feel at ease and at times even serene.

> "I like to fish when the weather is good," said Ivanildo. "I feel sorry when there is no more ice in the storage hold and we have to return ashore. When the nets have been set at night, we prepare a large can of coffee, make some sort of tent from a piece of canvas, put the gas lantern under it, and we play dominoes." "Don't you ever worry about your family?" I asked. "Eh," he answered, "they always find a way to survive when I'm gone, and in case of an emergency the boat owner or fish dealer will take care of them."

Thus, the drawbacks of boat fishing seen by canoe fishermen are interpreted as virtues by the boat fishermen. The boat owner with his powerful network can mobilize help more rapidly than any canoe owner. A boat owner who defaults on his responsibilities as an employer will soon be avoided by the best captains and fishermen. His reliability to the families must be impeccable for his business to succeed.

The economic situation of boat fishing is also not as bleak as the canoe fishermen have suggested, according to the boat fishermen. Boat fishermen can obtain goods on credit in local stores more easily because they are not independent producers with uncertain revenues but employed by boat owners who ultimately will assume most debts. Incurred costs can always be deducted little by little from the catch shares.

The credit advance system is central to the critique of boat fishing but its evaluation depends much upon who gives his opinion. High-producing fishermen praise the credit system as a stable source of income and as restraining spendthrift wives. They are able to save money because they receive a large lump sum of profits after each trip, while keeping down household expenses to a minimum by asking for a low credit advance. The *vale* facilitates saving and does not lead to indebtedness if one works hard. Canoe fishermen and poor boat fishermen share a different interpretation: the *vale* prevents families from starving but leads to indebtedness because the catch division is unfair and the fish prices are too low.

Finally, boat fishermen argue that motorboats are physically less demanding. Sometimes, canoes have to be rowed against strong winds and currents for hours, causing severe pains in the kidneys and the lower back. Furthermore, the greater geographical mobility of boats and the ability to preserve fish allow for longer and more productive fishing trips. Boats have better opportunities to make large catches than canoes and many boat fishermen claim that their annual revenues are higher than those of their northern colleagues. Such statements about income are, of course, never verified because the two groups seldom meet and even then will not exchange catch figures but only make vague allusions to their successes at sea. Only systematic data collection can establish the validity of the often inflated claims.

### *An Income Comparison Among Fishing Modes*

Any comparison of the productivity of several fishing methods is complex because the sample may not be representative. Crews change in composition in a year's time so that aggregate data must be compiled from different fishermen with different skills. Also, unusual seasonal variations may bias the figures unfavorably against a particular catching strategy. Even more problems arise when the data have been collected for only part of the year, as is the case here. Unlike agriculture with its seasonal harvests, fish production is daily or weekly. The recording of fish outputs demands constant attention because fishermen tend to forget what they caught a couple of days ago. Hence, I decided to collect data during five different, month-long periods: three in summer and two in winter. The canoes and boats represented in the sample are generally considered to be the most successful in Camurim.

The first period covers the poorest canoe fishing season.[7] The second period is the principal season for shrimp trawling, the third is the tourist season, the fourth has the highest incidence of canoe fishermen switching to boats, and the fifth is the time when snooks migrate between sea and river, and the catches of liners are best. The income figures are presented in average daily earnings during the five 31-day periods and have been rounded off to facilitate their comparison. Finally, to place the data in a cross-occupational perspective: the minimum monthly wage in the state of Bahia was Cr$21,000 (Cr$675 a day) in December 1982 and Cr$32,000 (Cr$1,000 a day) in May 1983. The minimum wage has an indirect influence on the fishing industry in that it is a standard by which boat owners determine the size of the credit advance. Indebted boat fishermen receive in the bad fishing season a monthly sum roughly equivalent to the minimum wage, while high-producing fishermen can ask for larger amounts. Despite great income variations, most fishermen earn more than the minimum wage, especially if the value of the fish taken for their personal consumption at home is added. Canoe fishermen consume at least Cr$10,000 (Cr$325 a day) a month and boat fishermen take around Cr$5,000 (Cr$160 a day) worth of fish home.

***Table 5.1***

Income Comparison of Boat Owners, Boat Fishermen, Canoe Owners, and Canoe Fishermen During Five One-Month Periods

| | *I* 9/21– 10/21/82 | *II* 11/19– 12/19/82 | *III* 1/10– 2/9/83 | *IV* 5/13– 6/12/83 | V 7/27– 8/26/83 |
|---|---|---|---|---|---|
| Canoes | Cr$ | Cr$ | Cr$ | Cr$ | Cr$ |
| Owner 1 | 1,325 | 1,740 | 3,390 | 1,500 | 3,100 |
| Fisherman 1 | 1,325 | 1,740 | 3,420 | | 2,900 |
| Owner 2 | 645 | | | 1,165 | 3,500 |
| Fisherman 2 | 580 | | | 835 | 1,600 |
| Owner 3 | | 1,515 | 2,325 | 1,315 | 935 |
| Trawler | | | | | |
| Owner | | 1,680 | | | |
| Fisherman 1 | | 840 | | | |
| Fisherman 2 | | 810 | | | |
| Netfishing Boat | | | | | |
| Owner | | 2,195 | 2,515 | | |
| Captain | | 610 | 615 | | |
| Fisherman 1 | | 645 | 645 | | |
| Fisherman 2 | | 160 | 710 | | |
| Liners | | | | | |
| Owner 1 | | 810 | 3,065 | 7,855 | 8,400 |
| Captain 1 | | 355 | 2,065 | 3,145 | 7,065 |
| Fisherman 1 | | 515 | 1,420 | | |
| Fisherman 2 | | 130 | 740 | 1,875 | 2,900 |
| Fisherman 3 | | 65 | 390 | 960 | 1,935 |
| Fisherman 4 | | −95 | 30 | | |
| Fisherman 5 | | | | 2.460 | 2,635 |
| Fisherman 6 | | | | 690 | |
| Fisherman 7 | | | | 645 | 1,235 |
| Owner 2 | | | | 1,020 | 10,435 |
| Captain 2 | | | | 440 | 3,035 |
| Fisherman 8 | | | | 335 | 2,600 |
| Fisherman 9 | | | | 270 | 1,735 |
| Fisherman 10 | | | | 190 | 1,300 |
| Fisherman 11 | | | | | 1,300 |
| Fisherman 12 | | | | 40 | 300 |

Before comparing the four fishing methods, I will discuss each category in detail. The first and the second canoe—in period I—fished with wide-mesh gill nets, but one canoe earned double the other, presumably because of greater luck and knowledge of the fishing grounds. The first canoe continued to fish with *caçoeiras* in the second and third period, while the third owner fished with fine-mesh *tainheiras* because he does not belong to a corporate group. In the second period, he earned Cr$61,000, but he had to pay Cr$14,000 to a mender to help repair his damaged nets, leaving him with only Cr$47,000 (Cr$1,515 a day) in profits. Periods IV and V demonstrate a comparison of three different fishing strategies. Owner 1 fished exclusively with caçoeiras for profitable but highly unpredictable snook. Owner 2 and fisherman 2 alternated between snook fishing and handlining. Owner 3 does not own wide-mesh nets and was obliged to fish with handlines, even though the fifth period is generally regarded as the worst time for shallow water handlining.

When the canoe fishermen were confronted with these figures, they expressed no surprise because they know that the highest incomes are earned in winter (periods IV and V) and in the tourist season (period III) when prices are raised by the high demand. Boat owners and boat fishermen were surprised. They had always assumed that the winter revenues of canoe fishermen would be low due to bad weather and the distance to the fishing grounds. "Still," a boat owner insisted, "this must have been an unusual year. Stay here next winter and you'll see that we earn much more." The statistical data did not overturn his beliefs but were interpreted in a way that confirmed his preconceived notions about the canoe fishing mode.

Trawling is considered a safe strategy during uncertain fishing seasons. Profits are never exceptional but the yields are always enough to pay for running costs and guarantee at least a minimum income. Period II is the best shrimp trawling season. I did not collect catch data in other periods because boats went trawling only sporadically.

The results of the medium-size netfishing boat show that summer is the worst season for boats. Occasional checks during the winter gave the strong impression that the revenues improved substantially when this particular crew went handlining.

Canoe fishermen and boat fishermen were not surprised by the income figures of liners because all know that summer is the poor, and winter the good, fishing season for large boats. During the second and third period, catches were so low that most boat fishermen got into debt, and some boat owners temporarily owed money to the wholesalers. Both kept receiving credit that was paid off only several months later. Canoe fishermen focused on these figures in their critique, but owners, captains, and a few high-producing fishermen countered that the high earnings in winter more than compensated for the bad summer fishing season. The indebted fishermen, however, considered their suspicions about the unfair catch division and the low fish prices confirmed.

It is impossible to draw any authoritative conclusions about the relative profitability of canoe fishing and boat fishing. In general, it is clear that the income variations between seasons and among boat fishermen are greater than those among canoe fishermen. Canoe fishermen do better in summer, but in winter some boat fishermen surpass the earnings of their northern colleagues. The income difference between capital owners and crews is much greater in boat fishing than in canoe fishing. Fishermen on small and medium-size boats are less likely to make large amounts of money but the chances of indebtedness are smaller than on liners. Unsuccessful fishermen on liners appear to be doomed to indebtedness if they do not succeed in switching to canoe fishing.

One cannot conclude that either canoe fishing or boat fishing is the economically rational choice because much depends on personal factors such as skill, ability, knowledge, persistence, intuition, luck, and factors such as catch division and market distribution. The average income of canoe fishermen varies much less than that of boat fishermen and confirms the popular notion that ambitious men tend to flock to boats. This state of affairs allows for the multiple interpretations of the pluriform economy and makes different persons justify their decisions in equally convincing terms.

## *Emergence of the Pluriform Fishing Economy*

Before the introduction of motorboats, northern and southern fishermen formed two relatively isolated groups, with little social interaction, who kept to their own fishing grounds but who had similar production modes. When boats began to appear in Camurim in the 1970s, the southern fishermen were particularly eager to abandon their canoes. River catches were declining, the trip upstream to the harbor could be arduous in a canoe and took far less time and effort in a boat. The demonstration effect of periodically sizable boat catches was convincing, and many fishermen were mesmerized by the possibility of ascending to the status of boat owner. Motorboats seemed somehow the wave of the future, of modernity, progress, and improvement. There even seemed to be an element of manliness involved, enhancing the virility of those daring enough to launch themselves into unknown waters at night.

After weighing the advantages of canoe fishing and boat fishing, the scale clearly tipped in favor of the boats. The short overnight trips to previously unreachable fishing grounds were very appealing. The fish were larger, catches better, earnings higher, and the companionship at sea was a pleasant change from the often lonely activity of canoe fishing. Many were sure that canoes had finally become obsolete and that now they could begin to explore the unfathomable and inexhaustible wealth of the ocean. Canoes were sold for almost nothing in surrounding villages and to some skeptical colleagues in the north who were less certain of the financial benefits. Most northern fishermen were also enticed by the new prospects. In many ways, boat fishing was similar to canoe fishing. The catches were sold immediately upon their delivery ashore and boat owners did not act differently toward crew members than did canoe owners. The fishermen could participate in community activities as before and much time could be spent with the family because boats left in the late afternoon and returned the next morning.

Boat owners were happy to receive skilled fishermen with a superior knowledge of the northern waters, and were careful to appease any tensions with southern fishermen through the generous flow of alcohol. The contacts did not precipitate a greater intimacy of their families, but the surge of fictive kinship relations indicated some approach between the two groups.

In the second half of the 1970s, when the Brazilian economy was hit hard by the global oil crisis, increasing numbers of fishermen began to have doubts about the economic advantages of boat fishing. The operating costs had been going up rapidly, the catches were declining, and to compensate for the lower productivity, boats stayed longer at more distant and dangerous locations. Northern boat fishermen were the first to notice the deteriorating conditions. They were able to compare their revenues with the income of neighbors who had never switched to boats. They realized that in volume the catches of coastal fishing were smaller, but that the average profits were higher. Furthermore, some began to resent their subordination to boat owners and their occasional indebtedness caused by a credit system that came into being when more time had to be spent at sea. Others were scolded by their wives for neglecting the family by their prolonged absence—an emotional starvation that was clearly not suffered in neighboring canoe fishing households.

The worsening conditions must not have gone unnoticed by the southern boat fishermen but the economic prospects of canoe fishing were unclear to them because there were insufficient canoe fishermen left in the neighborhood to make a balanced comparison. Southern boat fishermen mirrored themselves in the careers of poor colleagues who had become boat owners. They placed their hopes on lucky catches and rich friends.

The switch back to canoe fishing was not easy for the northern fishermen. Most of the men had sold their canoes in the first flush of enthusiasm as boat fishermen. Fortunately, they could embark as crew members in the canoes of neighbors, relatives, or friends, and thus save money to buy dugouts themselves. During this period, the pluriform fishing economy took shape and helped to formulate the distinction in the social identity of canoe fishermen and boat fishermen. Because of the contrasting experiences in the two production modes, their group identities became more pronounced and they became aware of aspects of life that in former times had been taken for granted. The social and personal identities of canoe fishermen and boat fishermen polarized, grew apart, and became consciously linked to the canoe or the boat as cultural foci.

### *Incentives and Impediments of Switching Between Modes*

Fifteen of the fifty canoe fishermen, ten from the north and five from the south, have not cut all their ties to boat fishing. They are in the privileged position of being able to switch back and forth between production modes. Around the time they want to embark, they hang out in the harbor and visit the bars frequented by boat owners and crews to look for vacancies. The men are always welcome because they have no accumulated debts and are known to work hard. The desire to fish on boats for a couple of weeks or months confirms the conception of southern owners and boat fishermen that canoe fishing is not as profitable as boat fishing. However, the motives for switching and especially the reasons to remain on boats are more complex than they seem.

Income is the most common motive for switching between fishing modes. Confronted with several weeks of poor catches in winter, a canoe fishermen feels that he is unable to provide for his family adequately. If the bad fishing season does not seem to be about to end, he decides to try his luck on a boat.

Domestic quarrels and conjugal conflicts can cause fishermen to abandon canoe fishing on the spur of the moment, without contemplating the impulse from a financial point of view. Amerino punished his wife by leaving little money at home and going on a long fishing trip. He wanted her to realize that he had always been a responsible husband, despite his temporary flaws. "I'm going to let her suffer like the wives of the indebted boat fishermen," he said angrily.

The third motive for switching lies in the public sphere. A canoe fisherman with a penchant for drinking will not optimize his productive efforts. He will regularly miss fishing trips and fail to mend his nets. His income declines, domestic tensions appear, and to escape the deteriorating economic and domestic situation, he seeks employment in the harbor. Boat fishing facilitates the drinking habit. Boat fishermen work only at sea and are free to do what they want after the catch has been unloaded. They can pass their time in bars and spend all earnings on rum because financial responsibility for the household is relegated to the employer through his credit guarantees.

Finally, a canoe fisherman may develop a desire to own a boat because he has made some contacts with tourists and local patrons. Suddenly, his lifestyle seems monotonous and his prestige insignificant. He redefines his position in the community through expectations awakened by the new asymmetrical relationships.

Usually, the motives for switching are preempted after some time. Canoe fishing yields improve, the wife reverts to a submissive role, the damaging domestic, economic, emotional, and physical consequences of unrestrained drinking are regretted, and owning a boat proves more difficult than was anticipated. Changed circumstances justify a return to canoe fishing. Although such sequences of action can be observed, behavior is seldom so straightforward. Human actions do not proceed along logical lines but are modified by the circumstances in which they are realized. These shifts in interests are common. They make a person steer clear of his initial objectives. Even if the fisherman is earning more than he used to, less income might be spent on household items because of an acquired drinking habit.

> Lucas lost sight of the motives that made him switch to Geraldo's boat. His earnings fell below his canoe fishing revenues but his consumption demands had changed and domestic financial concerns had become less important. After the first fishing trip, he visited his former colleagues at the beach and chatted a bit about his most recent experiences, but there was little real communication. They nodded their heads and kept on mending their nets. Lucas did not want to hang around the house when the canoes went fishing, so he joined the crew in the harbor to play a game of snooker, and they invited him to share a bottle of rum. At night, he went to bed late because the captain had decided to leave only after the weekend. The bar became the

> only outlet for his boredom. His wife was unaccustomed to his late-night outings and reproached him for his irresponsible behavior. He became angry at her lack of appreciation for his efforts, and became less committed to the household.

The main reason why a northern fisherman is not able to cope with the new circumstances for more than a few months is that he lives in a neighborhood surrounded by canoe fishermen. The canoe fishing households are an ever-present reflection on his former situation and the identity which emerged from it. Their life-styles represent what he has lost by temporarily suspending canoe fishing. Sooner or later, he realizes that with the passage of time he will become increasingly more of an outsider among the people with whom he identifies but fails to share any activities. They tolerate his absence as long as the reasons for switching persist, but he is expected to resume his place in the corporate group when the social and economic circumstances change. If not, the corporate group will exclude him from their redistributive support, the criteria for prestige among canoe fishermen will no longer apply to him, and he will be treated with some disdain as just another boat fisherman.

Initially, the former canoe fisherman still believes that his personal identity is not in conflict with his economic practice. He has not lost his desire for independence and autonomy, and feels out of place among the boat fishermen. Yet, sooner or later he is obliged to choose between different production modes, life-styles, peer groups, and social identities. The decision whether or not to return to canoe fishing is taken after an evaluation of the motives that caused the switch, the reasons that delayed it, and the new aspirations and goals that emerged from it. During this breakdown of the self, the fisherman may begin to complain about his forced subservience to the whims of the captain, the unfair catch division, hardships at sea, lack of respect from crew members, the threat of divorce, little time spent with his children, etc. The cultural centrality of the canoe has been invigorated through this reflection and has exposed the vexations of boat fishing.

Five of the 125 southern fishermen are canoe fishermen, although occasionally they also fish on boats. Owning a canoe at the riverfront is not the same as being a canoe fisherman in the north. The group is too small to give rise to a strong social identity reinforced through continued interaction and peer support. Southern fishermen are constantly confronted with jokes expressing the stereotypes held by boat fishermen about their northern colleagues. They become subjects of ridicule by association with canoes, not because their behavior resembles the stereotypes. The minority status, the lack of strong peer support, and the social isolation due to different labor schedules and consumption patterns make it difficult to persist in canoe fishing in a neighborhood where boat owners with their power, prestige, and superior living standard provide the role model for most men and adolescents. The pressure to conform is strong, and concessions in the public domain to prevent exclusion and stigmatization, such as bar attendance, gambling, and visits to brothels are tempting but fatal.

Southern canoe fishermen generally succumb to boat fishing rather than move to a more benign social environment where canoe fishing is the dominant mode of production. The social boundaries between north and south, and the problem of sea tenure, make it virtually impossible to move. A fisherman will have to find a group that will support him when his catches fail, and that will allow him to exploit their marine resources. Corporate groups are always afraid that others will take over their territories. The quick acceptance of new members might cause the invasion of newcomers to proliferate. Finally, southern canoe fishermen share many of the personal stereotypes about their northern colleagues and thus are not particularly interested in moving. Those who cannot sustain their way of life in the south will become absorbed in the boat fishing mode, despite their reservations about the benefits.

The move from boat fishing to canoe fishing has become more difficult than in the early days of the pluriform fishing economy. The most important reason is the fundamentally different outlook on life of canoe fishermen and boat fishermen. Boat fishermen are less predisposed toward a production mode that does not offer opportunities for economic mobility. Ambition has become part of their identity. Boat owners and high-producing fishermen will never switch to canoes because—for them—financial and social success have the highest priority. Three high-producing fishermen own canoes which they only use intermittently during the snook season in winter. Others sometimes borrow canoes but all of them regard permanent canoe fishing as a step down the social ladder into a static, lower-class position. Interestingly enough, switching high-producing fishermen go through the same role changes as canoe fishermen who switch to boats, only in a reverse and opposite order.

> During the first days, Alberto still hung out with his boat fishing colleagues, but he noticed his poor performance the next morning. He was too tired to row to the place he wanted to go to and returned after a few fruitless hours. That day he stayed at home in the evening and retired to bed early. He dissociated himself from his friends and during the following days began to seek the company of other southern canoe fishermen as they exchanged information and mended their nets. He could have become a permanent canoe fisherman were it not for his ambition and his negative assessment of canoe fishing in general. Alberto's self-perception clashed with the productive activities and life-style demanded by canoe fishing.

Around twenty poor fishermen have lost all faith in the boat fishing industry and would like to buy canoes and nets. Indebtedness is the most obvious impediment. They simply cannot save enough money to buy fishing gear. Southern boat fishermen cannot work their way up as fishing partners, as is common in the north. There are no vacancies on the few canoes in the harbor, while social boundaries inhibit them from going to the northern beaches.

Chronic alcohol consumption is also a serious problem. Impoverished boat fishermen who claim a desire to become canoe owners seem to have little predisposition and determination to achieve that goal. Poverty, indebtedness, and strained conjugal relations because of the family's low standard of living drive the

A quarrier prepares a mound of clay for a brickmaker.

A charcoalmaker is closing two furnaces with clay before firing them.

The fishing crew loads the beach seine into a large canoe while the shore crew (right) looks on.

A netmending party.

The harbor of Camurim. The large two-story building on the right is a former warehouse from the heydays of the cabotage of cocoa.

The Sunday market.

A fisherman awaiting departure. The mangroves that lace Rio Camurim are visible in the background.

A popular daytime bar.

A small weekend bar frequented by the elite.

Typical houses with their distinct façades (*pratibandas*). Note the porch that has been added to the house on the right.

A canoe fisherman surmounts the surf.

The launch continues . . .

men away from the house into the bars. The social environment does not stimulate them to switch. Furthermore, there is no encouragement from boat owners and high-producing fishermen to change production modes. Although poor fishermen hardly ever break even, they do supply much labor and produce a sizable share of the fish output. The industry could not operate without their manpower.

## *The Risks of Fishing*

"After creating the world, God asked the sea if it could offer a guaranty. The sea replied: 'One day yes, one day no.' God asked the river, and the river replied: 'One day yes, one day no.' God asked the mangrove, and the mangrove said: 'Every day.'" This tale was told to me by Celestino after a week of poor catches. "Fishing," he said, "is to venture (*aventurar*), to be in doubt about the outcome but with trust in God and one's ability to reduce this uncertainty." Iemanjá, the Goddess of the river and the sea, does not surrender her resources without resistance. Crabs and mollusks can be harvested daily from the mangroves, but fish have to be hunted without any guarantees of success. The sea personifies for Celestino an opponent who is continuously beaten but never conquered.

What sets fishing apart from other professions in Camurim is the uncertainty of production efforts and the fluctuations in earnings. Fishermen prefer this unpredictability to the security of a job ashore with a low fixed income. Fishermen experience fishing as an adventure. The adventure lies in its "uncertainty and freedom" (*incerteza e liberdade*). Each man and each crew tries to beat the odds of failure with different strategies and techniques. In this sense, the fisherman is given more leeway to make decisions than wage laborers are because there is no foolproof production method. Some believe in fishing at high tide, others at low tide, some at night or at dusk, with small hooks or large ones, with thin or thick fishing lines, at starboard or at portside. Aside from these smaller choices of skill and intuition, the pluriform fishing economy offers two major options to Camurimenses to deal with risk and uncertainty: boat fishing and canoe fishing.

The greatest preoccupation of canoe fishermen is to catch any fish at all. Coastal yields are small, and they will fish every day of the week to take advantage of the limited possibilities of success. Tides, winds, currents, salinity, the lunar cycle, spawning, migration, and other physical and biological factors which affect the presence of fish in shallow waters are of great concern. These natural vicissitudes enhance the cohesion of the men in corporate groups. They have developed information and distribution networks to spread the risks more evenly among them. They prefer these insecurities to the many manmade uncertainties experienced by boat fishermen.

The cooperation of boat owners and boat fishermen in dealing with the uncertainty of fishing takes a more hierarchical expression than among canoe fishermen. The captain-owner, or the captain in consultation with the boat owner, decides about the destination of a ten-day fishing trip when the weather is good. He navigates the motorboat to the best location known to him and the crew places its trust in his ability to land a large catch. This faith subsides when the circumstances are less than optimal. Unstable weather, strong winds and currents, an unfavorable phase of the moon, oncoming religious festivities, political events, or scheduled repairs may all shorten a fishing trip. Suddenly, the captain's opinion is no longer taken for granted, the routine of fishing has broken down, and the strategies at sea become evaluated in an open discussion ashore or aboard ship in which the captain and his crew present their options against a background of different economic contexts.

The crewmembers have no doubts about the captain's knowledge of the sea, but they realize that his personal motives may no longer coincide with their interests. How do they decide upon a particular fishing strategy? Should they take the greater risk of pursuing the most lucrative but also most scarce species; should they try to catch more abundant, lower-priced fish; or should they not leave the harbor at all? This problem is not simply solved by a cost/benefit analysis but makes fishermen, captains, and owners weigh consequences that reach far beyond the fishing trip into wider social concerns. This conflict is most apparent among the crews of large motorboats. The following description refers especially to these handlining crews, even though similar discursive confrontations occur in medium-size and small boats.

Boat owners assume that boats produce economic value which, paradoxically, can be maximized by waiting. Waiting becomes the deferment of work that is supposed to make capital more productive and improve the quality of the output.[8] They are convinced that it is better to wait for hours at a spot known to have high-priced fish in the hope that the fish will bite, then to go to an area where low-priced fish can be caught in abundance. Although the capture of scarce fish at distant locations demands a greater investment of time and capital per unit than that of cheap fish closer to shore, the quality and total revenue of the catch is believed to be superior in the long run. Large motorboats make accessible fishing grounds that are beyond the reach of smaller craft. Boat owners would have never invested in larger boats had they intended to use them close to shore. At times, the crew may fail to catch anything, but the boat owners believe that the few large catches of high-quality fish will more than compensate for a string of poor catches that will hardly pay for the costs of these unsuccessful fishing trips.

Boat owners do not care about the operating costs because these are always deducted from the catch and thus are usually paid for, nor are they as much interested in short-term profit maximization. They not only believe that ultimately a persistent search for high-quality fish will be more profitable, but they fear that their relationship with wholesalers will deteriorate if their boats do not bring in the best fish species. Dealers will deliver their supplies of ice, fuel, and bait more quickly to the boats which bring in the most desired species. Crews that make shorter trips and catch lower-quality fish will be the last to sail.

Yet, there is a dilemma: fishermen cannot afford to wait indefinitely for the maximum return and accept the increased risk of not catching anything at all because they have families to feed, households to maintain, and career interests to look after. To solve this subsistence problem, boat owners give fishermen credit to bridge the time and financial gap between labor spent and output produced.

High-producing fishermen prefer to go on trips with reduced

running costs and the possibility of a sizeable catch of medium-quality fish, instead of spending many hours sailing to fishing grounds with dubious results. They want to maximize the profit in the short run and reduce the level of uncertainty by fishing on more abundant species to guarantee a more even flow of income. High-producing fishermen are concerned that by taking too great risks they will not always have enough cash on hand to maintain their valuable social contacts with potential patrons.

Word that the fish are not biting discourages low-producing fishermen from embarking on long trips, because they are afraid of incurring unpayable operating costs. They prefer to stay ashore or, at most, make overnight trips to avoid the high expense of food, ice, and fuel. They are satisfied with catching only low-quality species of fish, which are not wanted by the wholesalers but that can always be sold to the poor people of Camurim. They are less concerned about their immediate domestic needs because the credit system allows them to wait for better days. It is therefore in their interest to paralyze the production process altogether when the fishing conditions are not optimal. Strangely enough, waiting—the temporary suspension of labor—generates income for the low-producing fishermen. Capital and labor have not become antagonistic sources of economic value, as they are for boat owners and high-producing fishermen, but have become each other's inversion in which the nonproductive use of capital sustains the reproduction of labor.

The ways in which this conflict is worked out depends greatly on the relation of authority between owner and crew. An absentee boat owner has no direct control over the decisions at sea but can dismiss a captain who comes ashore too often with low-quality fish. A captain-owner can, of course, impose his will on the crew but he will lose his best fishermen if he does not seek a compromise. How each disagreement is resolved cannot be ascertained in advance. What is important is that the dispute lingers on and might erupt again, only to be decided otherwise when the balance of power shifts, because each actor situates the practice of fishing in a different economic context of personal significance and interest.[9]

The reasons why fishermen choose one or the other fishing

strategy or production mode are not restricted to the economic organization, and not even to the economic practices they perceive. Economic practices are related to social practices considered to be of crucial importance by the fishermen. These social practices give rise to the economy, just as the economy gives rise to them. Constructed around cultural foci, these social practices relate to economic practices in a dialectic manner, with one cultural focus instilling meaning in, and deriving meaning from, the other.

# 6
# Sea, House, and Street

> The first time I became personally aware of the many subtle tabus in Camurim concerning the transition between the street and the house was when I bought a large papaya several months after I had begun my fieldwork. On this particular Sunday morning in October 1982, my wife and I were returning from the market when I spotted a nice papaya on display in a windowsill. I approached the house and—through the window—asked the woman inside for its price. As I tried to hand her the money and take the fruit, she shook her head and took the papaya from my hands. I didn't understand her reaction and thought she wanted more money. I asked her again if I could buy it. She said yes. I tried to give her the money and once more she refused to give me the fruit. Then she asked me to step inside the small front room and finally handed me the papaya. I asked her why she hadn't given it to me before. "because it gives *atraso*," she replied, "it brings bad luck." The window is not a passageway between the house and the street. People and things enter and leave the house through the front door.[1]

This chapter describes similar expressions of habitus which structure people's practices but which are seldom the subject of discursive dispute. I will demonstrate that the division of Camurim's social universe into economic, domestic, and public domains is not an ethnographic model but is firmly grounded in the social practices of the fishermen. I will outline the spatial and material boundaries of this social universe, specify the transitions between the domains, describe the domestic world which first socializes people into these divisions, and analyze how these hegemonic, habitual ways organize the practices of the fishermen on the domestic and societal level.

Bourdieu defines habitus as a confluency of agency and structure, an enactment of "systems of durable, transposable *dispositions*" (1977:72). "The habitus is not only a structuring structure, which organizes practices and the perception of practices, but also a structured structure: the principle of division into logical classes which organizes the perception of the social world is itself the product of internalization of the division into social classes" (Bourdieu 1984:170). Habitus simultaneously generates and is generated by social classifications reproduced in practice. Bourdieu's habitus is similar to Sahlins' structure of conjuncture (1985:vii–xvii), Giddens' structural duality (1979:216–222), Gramsci's cultural hegemony (1971), and Foucault's deep force relations (1978:92–97). The habitus is so close to people's cultural being that they are unaware of its structuration of their everyday behavior. Material and spatial objectifications of cultural meaning, such as implements, consumption goods, dress, urban designs and architecture, are particularly important instances of habitus because they tend to persist longer in time than interactional manifestations, and may even become cultural foci which structure and orient whole clusters of social practice.

The sociospatial organization of Camurim reflects the social stratification of the community, reinforces status distinctions and class affiliations, and inculcates a division of the social world into a domestic, public, and economic domain. A child in Camurim is socialized into this particular social universe as it ventures out into the streets and roams the market, the main square and the center of town (cf. Gilmore 1977; Richardson 1982). This process of inculcation does not involve any conscious memorization because "schemes are able to pass from practice to practice without going through discourse or consciousness" (Bourdieu (1977:87).[2] A child grows into these cultural divisions and practices. These objectivations of routine practice are apparent from the different ways in which fishermen pass from one social domain to another. All fishermen move back and forth between the house and the street, but canoe fishermen go directly from the sea to the house, while boat owners and boat fishermen pass betweeen the sea and the house via the public domain. Social domains are mediated by transitional spaces in which persons undergo status transformations when they enter another social context. Although all men participate in the three domains, they do not pass through the same transitions and status changes.

The house in Camurim is one of the most important objectifications of habitus which "through the intermediary of the divisions and hierarchies it sets up between things, persons, and practices . . . continuously inculcates and reinforces the taxonomic principles underlying all the arbitrary provisions of this culture" (Bourdieu 1977:89; cf. also Bourdieu 1973).[3] A Brazilian child acquires its first basic conception of social interaction in the home, and reproduces this cultural interpretation when it sets up its own household.[4] The average house in Camurim has three main areas, each characterized by specialized domestic practices: the front room, the kitchen, and the bedrooms. The sheer existence of these rooms, the social functions alloted to them in Brazilian culture, and their importance in childhood and in socialization create a need for their embodiment in social interaction. This domestic sociospatial organization corresponds to the division of Camurim society into three social domains. The kitchen is the area where the inhabitants reproduce themselves economically as a household. Household relations are expressed in material and emotional ways by sharing meals, affections, worries, and hardships. The parental bedroom is the area of conception, sexual intercourse, and physical recuperation. Here, in a very literal way, husband and wife shed their clothes and retreat within the intimacy of their bodies, just as they withdraw from society into the security of the domestic world. The front room is the space in which the family relates to the outside world, receives guests, and presents itself to the community with the display of its most precious status symbols. In this chapter, I will describe the material and spatial definition of the three areas of the house and how they relate to the three social domains of society. In the next chapter, I will analyze the social dimension, elaborate on the correspondence between societal and domestic interaction, and indicate how the content of the social interaction in the three domains varies according to the social functions and role relations of the inhabitants.

Aside from the location of a house in Camurim in the sociospatial organization of town, its exterior has a great symbolic importance. The façade and the front porch are two architectural features which communicate to the outside world the social status of the inhabitants. This material expression of social differentiation is especially pronounced among the canoe fishermen. In the last section of this chapter, I will discuss how canoe fishermen acquire social prestige with these architectural embellishments, through the luxury goods they purchase, and the standard of living they maintain in the household. Canoe fishermen give much symbolic significance to the rooms of the house and the belongings which define their social context. Although boat owners and boat fishermen occupy similar rooms and possess the same objects, they do not give them the same centrality in their social practices.

## *Social Space and Status in Camurim*

Camurim was laid out in a checkerboard pattern in 1764. This urban plan so common in Spanish America was introduced to Brazil in the eighteenth century by military engineers (Hardoy 1978). The importance of the river as the focus of communication with the outside world prevented the development of a central space that could concentrate the administrative, political, religious, and economic institutions along its sides. The wooden church, the town hall, the jail, the courthouse, and the military post were strung along the riverbank as palisades of authority facing the strangers arriving from the sea and beyond.

The erection of a large church, at the end of the nineteenth century, outside the old nucleus marked the beginning of a reorientation of social space. The mere presence of this church consecrated the old, poor neighborhood. The landed elite, always keen on remaining close to the sacred and the pure, bought out the fishermen living in its vicinity and constructed large mansions in sight of the church. Each consecutive municipal administration added its touch to the relocation of religious and secular authority.

The most significant change occurred when the main street was paved and its midsection set aside for the construction of public buildings. Within a decade most public services were located along the main street (see figure 6.1) This city plan clearly reflects the social and economic development of Camurim since the 1960s. Fluvial and maritime transport have disappeared, the old streets in the harbor district have become too narrow for trucks to pass, and the topography of Camurim has followed the transformation of its social organization. The relations among the significant social groups of Camurim have found an architectural and spatial expression that reinforces their social and cultural distinctions (cf. Gilmore 1977).

The main street has not yet attained its final form. The prefect who won the 1982 municipal elections wants to relocate the bus stop from its current location near a bar to a new terminal that will be constructed at the marketplace. He dislikes the sight of the bustling Sunday market with entrails and pigs' heads hanging from the stalls of butchers, the smell of overripe plantains, exotic spices, and herbal medicines. The roguish behavior of the teenage carters and the noise of stray dogs fighting over the discarded bones, all in the shadow of City Hall, seem undignified to him. The prefect wants to move the market away from the center to a location in the poorest part of town. He also wants to install a public sound system along the main street to broadcast regional radio programs from sunrise till sunset. This installation will finally give Camurim the urban atmosphere so characteristic of most towns in the cocoa region.[5]

The wealthiest inhabitants of Camurim live around the church in well kept two-story houses with glass windows and ornamental grates. This neighborhood has paved sidewalks, very few stores, and only expensive bars and nightclubs. Their former houses near the sea were bought by the members of a younger generation of descendants who left Camurim to receive a higher education, settled in Rio de Janeiro, São Paulo, Vitória, and Salvador, and

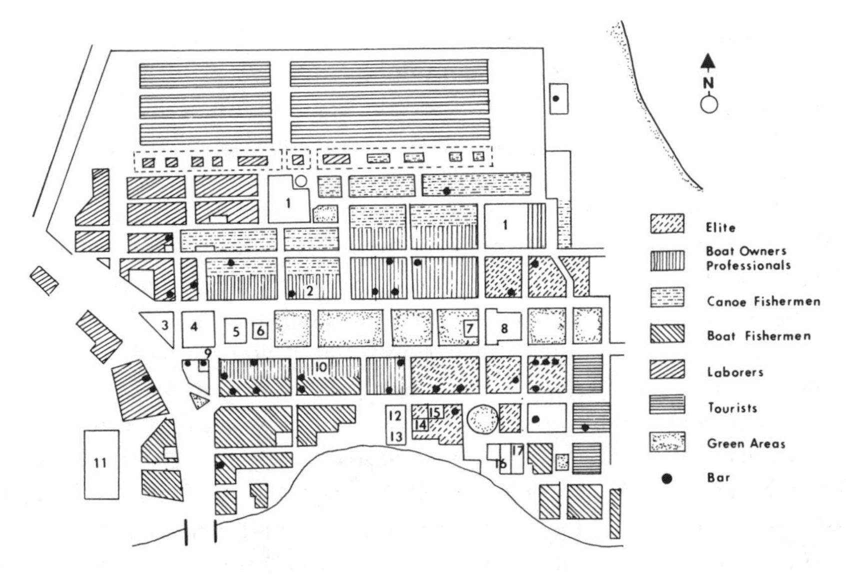


***Figure 6.1***
Sociospatial Map of Camurim

1 School
2 Church of Jehovah
3 Gasoline station
4 Marketplace
5 City Hall
6 Bank
7 Telephone company
8 Roman Catholic Church
9 Bus stop
10 Electricity company
11 Cemetery
12 Police station
13 Jail
14 Post Office
15 Water company
16 Bank
17 Cinema

return to Camurim only during the summer holidays. Boat owners, public officials, bank employees, and other middle-class professionals live along the main street near the bank and the City Hall. The one-story houses are smaller than those of the elite but have a modern façade and a front porch that mark the relative affluence of their inhabitants.

The main street is avoided by all who do not have business there. Fishermen who visit another part of town hurry across or take a detour by way of the beach. Most likely, they feel embarassed in their working clothes as they walk with self-conscious awkwardness under the scrutinizing eyes of the more sophisticated residents. The entire community fills the main street only on market day and during special occasions such as carnival, processions, parades, political campaigns, and secular and religious celebrations.

Canoe fishermen live in the northern part of town near the beach. The small one- and two-bedroom houses with colorful façades, Dutch doors, and wooden shutters show signs of remodeling. The old harbor quarter looks poorer. Several boat fishermen and their families still live in wooden houses, façades have lost their paint, and there are few television antennas on the roofs. Broken cobblestones enhance the timeworn appearance of what used to be a more prosperous past.

The poorest townspeople live at the entrance to Camurim. Day laborers, streetcleaners, sawmill workers, and brickmakers occupy the simple wooden and wattle and daub houses, some still with palmcovered roofs.[6] Many houses do not have electricity or latrines. The unpaved streets turn into mud in winter and give a desolate impression intensified by the children hovering in tattered blankets on the doorsteps. The neighborhood is such an eyesore to the elite that the last few city administrations have been carrying out an urban plan that removes an entire row of poor houses (represented as dotted rectangles on figure 6.1) from the northernmost part of town. The area is adjacent to Camurim Novo (New Camurim), the six blocks of villas owned by tourists. Numerous houses have already been demolished. The remaining residents cannot remodel their places and the electricity company has been forbidden to supply them with energy. The prefect wants to lay out a promenade with benches, flowers, and plants as a green buffer zone between Camurim and Camurim Novo.

Camurim Novo was planned between 1972 and 1976 by a prefect who saw that neighboring towns were attracting rich tourists while Camurim, with its equally beautiful beaches, remained undiscovered. He confiscated the properties at the current location of Camurim Novo and divided the area into building sites that were free to anyone who would start constructing a house within one year. Fishermen were barred from claiming lots, but many members of Camurim's elite took advantage of this investment opportunity to construct spacious houses and rent them out in summer or sell them with large profits.

Camurim Novo is still unpaved but it has electricity and running water. Each elegant white-washed villa is set in the center of a lawn surrounded by a waist-high wall. The neighborhood is purely residential and lacks both stores and bars. The houses are only occupied from December through February or March. Many summer residents come from the state of Minas Gerais. They do not have blood ties with the townspeople but maintain friendly relations with the persons who keep their houses clean. The tourists do not want to become involved in the social life of the community but spend most of their time on the beach and attend parties at the homes of other tourists.

The Brazilian preoccupation with living in a neighborhood that corresponds to one's social position, or even better, living slightly above one's status has been analyzed well by Velho (1975) for Rio de Janeiro and can also be found in Camurim. Walking through Camurim every day and observing the street scenes, I became increasingly aware of how each neighborhood reflected the socioeconomic status and lifestyle of its residents. Watching the people of Camurim participate in the street life of different neighborhoods, I sensed that they behaved routinely in ways I tried to grasp consciously.[7]

The neighborhood of canoe owners and canoe fishermen with its unpaved sidewalks always had an untidy look but the intense social activity of the inhabitants gave me a cozy impression. When I returned with the men from the beach in the late afternoon, they would leisurely mend their nets while exchanging hints about the next day's weather and the whereabouts of large schools of fish. They kept an eye on the children who played on the sidewalk with clay marbles but who sometimes raced out on the street in the excitement of the game. A woman with a toddler in her lap shouted at the men to keep a closer watch on the children while some older men seated in front of their houses observed the spectacle with amusement. The background was often filled with the sound of a television set as a group of teenagers watched a program through an open window. I felt most at ease in this neighborhood as I could see the family life on the street. People walked in and out of each other's homes and little girls trotted from group to group to call their fathers and brothers for dinner.

The center of town was usually empty, although I could hear children playing in the walled backyards or on the front porch under the watchful eye of a domestic servant. When I visited these homes during the week, I would often be attended by one of the servants. The women would be visiting a friend, while the men were away on business or at their estates. The movement of people was most intense around the public buildings, but the streets themselves served only as thoroughfares between destinations.

If the center of town had an atmosphere of authority and reserve, and the northern neighborhood exuded sociability and hospitality, then the harbor quarter gave me the feeling that there was something incomplete and contradictory about the street scene. Unlike the other neighborhoods, women and children dominated the street because the men were either at sea, in a bar, or at the harbor. This was unusual because, in Brazil, the street is a male space in contrast to the female space of the home (DaMatta 1987; Freyre 1961:30–66). During the entire day and into the evening, women passed back and forth between their house and the street, performing domestic tasks that were normally done in the backyard. They took care of each other's children playing in the street and warned them to stay away from the river. This street life made it difficult for me to spend much time with the women when their husbands were away, as I felt that my presence might give rise to unwelcome gossip.

The poor western neighborhood of day laborers and salaried workers was strikingly empty during the week. All able-bodied men, women, and adolescents were out at work, leaving the streets with abandon. Sometimes, the upper half of a Dutch door would open as I strolled by and an old woman would peer from the crack only to slam it shut seconds later. Unlike households in other neighborhoods, the old people lived with their children in extended households and took care of the grandchildren as the parents left for work. Poverty compelled these extended families to pool their meager income from small retirement benefits, minimum salaries, and occasional earnings from washing clothes and cleaning shellfish.

## *The Transition Among Social Domains*

The opposition between the house and the street is a basic feature of Brazilian culture which dichotomizes society into different social domains based on other interactional, ideological, hierarchical, and structural premises, according to DaMatta (1981:70–74, 1987:39–54) and Freyre (1961:30–66).[8] The street is a public domain of unexpected events, and of actors with no readily identifiable status. It is a predominantly male territory with hierarchical relationships, with individuals and social classes engaged in a never-ending struggle for supremacy and power. The house, on the other hand, is a domestic domain of members related by blood and marriage, with precise rights and obligations organized along age and gender (cf. Woortmann 1982).

The Brazilian anthropologist Roberto DaMatta (1981:70) defines the house and the street as the binary extremes of a continuum.[9] The street is a passionate, unrestrained, deceitful, and dangerous place of the masses with conflict, change, and action. The house is an authoritarian, restrained, intimate, calm, and safe place of tradition, harmony, loyalty, and leisure. DaMatta (1982:30) regards the house and the family as the opposites of the street and the world, where society manifests itself through the impersonal social forms of laws and bureaucracies. People work in the street and rest in the house. DaMatta has used this binary model successfully in his analyses of carnivals, religious processions, military parades, and street violence in Brazil, but the dichotomy seems to fall short when applied beyond the ritual realm in the everyday life of Camurim.

First, fishermen distinguish between the house, the street, and the sea (*a casa a rua e o mar*). The street must not be confounded with the workplace.[10] The economic domain has a strictly organized division of labor with well-defined statuses, chains of command, and distribution patterns. The street, on the other hand, is a volatile social universe in which the social divisions of the community become visible and individuals compete for prestige, power, resources, and relationships which indirectly bear upon their domestic and economic domains. A man's stature in the public domain is primarily determined by the social position of his peer group, his political influence, and the deference and demeanor with which he is treated. In sum, at sea he works, at home he rests, and in the street he manipulates.

Second, the degree of participation in the three domains varies among the social groups. Ranked according to the quality of their households, the home is the principal focus of interest for canoe fishermen. For boat owners and boat fishermen, the social life of the street is paramount over that of the house. In public they can display and obtain prestige, and meet people who can advance their social and economic careers. Boat owners and boat fishermen do not manifest their social status in the house, but in bars, brothels, and other public places.

Third, the house is a cultural focus for canoe fishermen, while the street is less important to their identity and interests in life. They have a strong awareness of family life as a domestic world which has to be sustained to give their lives meaning. The house lacks this wholesomeness for boat owners and boat fishermen because the family is less crucial to their public objectives. The bar, as the culmination of the public domain, is a cultural focus that represents their desires and wishes, and motivates them to work. Their domestic life is far more restricted in time and emotion than is that of the canoe fishermen.

Fourth, DaMatta seems to suggest that the meanings of the street and the house are unambiguous and do not vary across age, gender, class, or occupation. However, as can be deduced from the above and the three chapters on economic practice, boat owners, boat fishermen, and canoe fishermen define, evaluate, and interpret the transactional content and the interrelations of the three social domains differently, even though they may cooperate routinely in these social settings.

The house, the sea, and the street are separated by social and spatial boundaries that impede the free transition from one context into another. The most common features are speech patterns, vocabulary, interactional behavior, clothes, and especially the presence of transitional spaces. Intermediary spaces and boundary markers prevent persons from violating new contexts by an abrupt presence but allow for an entry and exit that mediate the social boundaries. The importance of these boundaries is apparent from an analysis of the differing ways in which canoe fishermen and boat fishermen go from one social domain to another.

Canoe fishermen and boat fishermen who enter or leave a house through the front door pass between the sidewalk (*calcada*) and the front room (*sala*). The sidewalk is an ambiguous space. Although part of the street, it can be incorporated into the house. When the men are at sea, women like to extend large mats of plaited straw on the sidewalk for the children to play on while they sit on the doorstep cleaning a bowl of rice or beans. In Camurim, as in other parts of Brazil (DaMatta 1981:74), men sometimes place chairs on the sidewalk after a day's work and chat with neighbors and passersby. Another indication of the sidewalk as an intermediate space is that people walk on the street instead of on the sidewalk. Persons will only pause on a sidewalk when they have the intention of talking to the people in the adjoining house.[11]

A rich folklore used to surround the transition between the house and the street, a folklore that was especially centered on doors and windows (Cascudo 1954:161–162, 514). Many of these customs are no longer practiced but those that persist indicate the existence of a social, physical, and symbolic barrier between the house and the street. The street is seen as polluted, as a source of evil, as a place of danger and conflict that may harm the members of a household if its bad influences are allowed to penetrate. People paint a crude five-pointed star (Cruz-de-Salomão) or the initials JMJ (Jesus, Mary, Joseph) on the front door to ward off the evil eyes of passersby. The initials represent the Holy Family and indicate the sacredness of the house, as well as the harmony and intimate relatedness of the inhabitants.

Celestino and Cabeludo made sure to wipe their feet on the decorated doormate of Casimiro's house when they entered. They made little effort to scrape the wet soil off their thongs

> but they wanted at least to make a symbolic gesture indicating that they brought no evil to the house and had left all hostile feelings behind. A doormat was conspicuously absent at the rear entrance where dirt and sand were constantly carried into the house as the children chased each other around the backyard and ran in and out of the kitchen, hiding under the table or in the bathroom. Casimiro's twelve-year old daughter, Aristeia, was sweeping in the front room. She never swept the house from the back towards the front, afraid that otherwise the happiness of the house would be thrown out with the dirt. For the same reason, Celestino and Cabeludo left the house later that evening through the same door through which they had entered. If they had not, unhappiness might befall the family.

The front door is the principal passageway into and out of the house, while the glassless window with its wooden shutters seem to exist only to observe the street activity or to let in some light and fresh air. The function of the door is to protect the inhabitants from the unwanted intrusion of outsiders. Its symbolic meaning is to divide the private from the public and to communicate the existence of this boundary to others. Outsiders must respect the integrity of the inhabitants and adhere to the practices and appropriate forms of conversation of the house. Camurimenses tend to turn away from closed doors and windows, signs that the inhabitants do not want to be disturbed. A closed door and an open window indicate that the family is in the backyard and can be visited by friends and relatives. During most of the daytime, the lower half of the Dutch door is closed to prevent dogs from entering and children from leaving, while the upper half of the door and the window are left open.

The mode of transition between the house and the sea is less uniform. Canoe fishermen overcome the change from work to home in a different manner than boat owners and boat fishermen do. For canoe fishermen, the beach, the backyard (*quintal, fundo*), and the side entrance (*passeio*) are intermediate spaces between the house and the sea. The beach is a male territory where canoe fishermen prepare themselves for fishing trips or where they rest briefly after arriving ashore. They never enter the house through its front door when they return from the sea but take the narrow passageway that connects the sidewalk with the backyard. It is not just their concern to avoid dirtying the house with fishing gear which makes canoe fishermen take the side entrance, because a bicycle, for instance, that has collected mud and grime off the streets is wheeled freely through the front door and kept in the front room. Instead, it is the symbolic separation of the house and the sea that is at stake here. The bicycle is a prestige object used in the public domain and expresses the social status of its owner. The front door is used when the canoe fisherman enters or leaves the public domain, the side entrance when his action concerns labor.

The backyard is an intermediate space which belongs to the house but where ambiguous activities that come close to being classified as street behavior may be done. Francisco's family uses an open latrine enclosed with plaited palmleaves in a corner of the backyard farthest away from the house, while the youngest children sometimes relieve themselves in the street when they awake. Garbage is collected in the backyard and, via the side entrance, dumped into the street. Francisco always washes himself in a fenced-off bathing area of the backyard after a fishing trip, even though he has a shower next to the kitchen. He changes his clothes and leaves the house through the front door. In the morning, he gets into his working clothes and leaves the house through the side passageway.

Boat owners and boat fishermen also enter and leave the house via the side entrance when they come from or go to work. However, the transition to and from the sea is more elaborate. Boat owners and boat fishermen go first to a daytime bar—before they go home—upon their arrival ashore. In the daytime bar, the public status of boat owners and boat fishermen is redefined before they enter the community. Their status undergoes small shifts according to the standing of their debts and credits. Without this information, Clovis would not know how to interact with his wife. She had been asking for weeks to buy a pressure cooker but Clovis kept putting it off with the promise that soon the good fishing season would start. Now that the weather has improved and he has made a good catch, he can ask the boat owner for a larger *vale* and allow his wife to make the first down payment. His friend, Isidoro, has heard that there is a small boat for sale but he first has to know his net earnings before approaching his patron. All public encounters would be problematic if the fishermen's ambiguous status would not be resolved. Boat owners and boat fishermen only go directly home if they arrive late at night or if they are ill or injured. In the first case, they appear at the bar the following morning; and in the latter case, their illness overrides routine status concerns.

> Life in Camurim had followed its unpredictable course while the crew was at sea during the second week of November 1982. João the bar owner, Geraldo the boat owner, the fish dealer, and two regular customers gave Isidoro and his crew the latest news. "You know who died?" began João. Isidoro shook his head. "Guma. Eh, Guma has died. We don't really know what happened but some woman found the body near the sawmill this Monday. Zé Espalhado said that he had met Guma late Sunday evening, completely drunk, as usual. He must have gone to the boat to take a nap, and fallen overboard. The incoming tide carried him upstream." The men paused for a minute, and then resumed their conversation. "I have good news," said the fish dealer gleefully as he raised his glass in the air, "we're going to raise the price of fish fifty cruzeiros!" The reaction of the crew was noncommittal. They had been arguing for weeks that they could not make ends meet, so they had been expecting a price raise. Geraldo broke the uncomfortable silence with the news of his nephew's upcoming wedding, and the political machinations of the prefect who made his illiterate brother-in-law the county's new land surveyor. All these persons deserved to be treated differently than they had been before the fishing trip. Such knowledge of changes in their relative social standing would prevent embarrassments and allowed Geraldo and his companions to act with appropriate care.

The daytime bar is also the place where boat fishermen overcome the abrupt change from a total to an open institution, as I will describe in detail in chapter 8. They make a transition from an imposed and, at times, tense association at sea to a voluntary and often desired dissociation ashore. The crew dissolves in the public domain and its members join separate groups. As members of the crew, they are all fishermen in the eyes of the community but, once ashore, they want to deny this common identity and emphasize their social differentiation.

When the men go fishing, they pass first through the bar, discuss the boat's supplies, and those who had not yet received their advance rush home to leave the household money. The men tease each other about their conduct ashore and the low-producing fishermen promise to improve their output. Alcohol is not consumed but the general atmosphere breathes the same spirit of equality that is felt when the boat arrives. They have to leave their public status behind, forget about the problems at home, and assume the responsibilities and attitudes of the labor relations aboard ship.

Entry into the public domain from the sea is not unconditional. If boat fishermen have to make a necessary purchase, they will wait outside the store on the sidewalk and ask the shopkeeper to bring the goods to them. Canoe fishermen will never enter a shop or bar after a fishing trip. On their way home, they avoid the main streets whenever possible. Amerino walks through alleys and unpaved streets inhabited by the poorer people of Camurim to get to his house. He lugs the heavy load of nets on his back, tired from hours of rowing. A shortcut through his own street could save Amerino a quarter of the distance but he does not want to engage in any conversation in his present condition. Canoe fishermen go directly home after arriving at the beach. Their fishing trips are short and do not have the characteristics of a total institution. Corporate groups do not dissolve in public, and the public status of a canoe fisherman is defined through the domestic sphere. He does not receive earnings from the daily catch in the presence of colleagues but at home, and indicators of public status are found at home, not in the street. A bad catch will not affect his public status or the demeanor of colleagues toward him because they are aware of his overall economic performance and are familiar with his domestic standard of living.

The spatial and material demarcations of the sea, the house, and the street, the existence of transitional spaces between them, and the systematic trajectories of fishermen as they pass from one sphere to another accentuate and establish clear boundaries to the three principal sociospatial domains of Camurim society. By taking certain routes and avoiding others, the boat fishermen and canoe fishermen express their preference for different social transitions and status transformations. The house is the pivotal domain for canoe fishermen, both in the sense of connecting the other two spheres and as a focus of their lives, while the public domain occupies that position for boat fishermen. Still, despite these differences, houses of the boat fishermen and canoe fishermen are surprisingly similar in construction and design, and function prominently in reproducing a particular conception of social interaction.[12]

## *Social and Spatial Organization of the House*

The architectural design of the typical house in Camurim has changed little for over two centuries. The culturally proper way to build a house dates back to instructions given in 1755 by the Portuguese Crown to the authorities of Camurim.

> Every house of every village must have a front of 50 *palmos* [11 meters], a depth of 35 palmos [7.7 meters], be composed of a front room with a door and two windows toward the street, a bedroom for the parents, another one for the children, a pantry and a kitchen, and a backyard that must be 80 palmos [17.6 meters] long and 50 palmos [11 meters] wide, corresponding to the confines of the house. All houses must have the same front, height, doors, and windows, and in order to avoid deviation from these regulations, the said Minister will help with the foundation and construction of a few houses, leaving the rest outlined. (*Revista Trimensal* 1896:537; my translation)

What is so remarkable about these instructions is that they are a conscious effort to institutionalize in colonial society a spatial and social organization of the house that was just emerging in eighteenth-century middle-class European society. The Europeans of the seventeenth-century still lived in general-purpose houses. "In the same rooms where they ate, people slept, danced, worked, and received visitors" (Ariès 1962:394).

> In the eighteenth century, the family began to hold society at a distance, to push it back beyond a steadily extending zone of private life. The organization of the house altered in conformity with this new desire to keep the world at bay. It became the modern type of house, with rooms which were independent because they opened on to a corridor. . . . There were no longer beds all over the house. The beds were confined to the bedrooms. . . . This specialization of the rooms, in the middle class and nobility to begin with, was certainly one of the greatest changes in everyday life. (Ariès 1962:398–399).[13]

With the exception of some nineteenth-century mansions and the large, modern residences constructed by tourists and rich townspeople, most houses in Camurim—including those of boat owners—have a front room, two bedrooms, a kitchen, a pantry, and a backyard. The houses are 5 to 6 meters wide, 8 to 10 meters long, and have Dutch doors and one or two windows at the front and at the back (see figure 6.2). Poor boat fishermen and many canoe fishermen live in wooden or wattle and daub houses with tile roofs.[14] All other houses are made of brick. Houses are without lofts and the inner walls do not reach to the roof. The circulation of air in the ceilingless rooms keeps the houses cool but also reduces the privacy of the inhabitants, especially because the bedrooms do not have doors but flimsy plastic curtains. When the kitchen becomes too small because of the purchase of a gas stove and refrigerator, a veranda—which stands in open connection to the backyard—is added.[15]

The division of the house into three areas (front room, bedrooms, and kitchen/veranda) and the tabus that surround their use regulate the social behavior and social relationships of the family members.[16] This domestic habitus determines which activities are proper in the house, where they should be exercised, and which should be carried out elsewhere. The goods in the house further reinforce the social definition of the three social spaces.

The kitchen/veranda is the emotional and interactional center of the household. The veranda has a simple lean-to roof under which the large wooden kitchen table and some small stools are placed. The family spends most of its time together in this area of the house. Here, the household, the nuclear family, and the

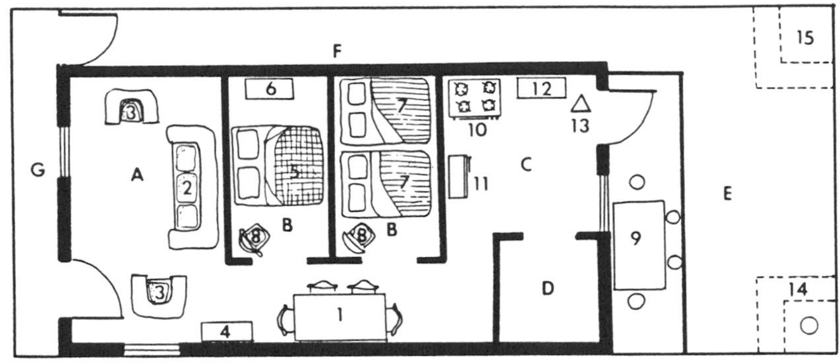


***Figure 6.2***
Floor Plan of a Typical House in Camurim

References of Diagram

A Front room
- 1 table and chairs
- 2 couch
- 3 armchair
- 4 cabinet

B Bedroom
- 5 conjugal bed
- 6 wardrobe
- 7 children's bed
- 8 chair

C Kitchen and veranda
- 9 kitchen table and stools
- 10 gas stove
- 11 refrigerator
- 12 cupboard
- 13 storage rack

D Pantry

E Backyard
- 14 bathing area
- 15 open outdoor latrine

F Side passageway

G Sidewalk

domestic relations are realized in practice. Canoe fishermen state that fishing becomes visible in the kitchen. How the household is reproduced through the consumption of food and through the emotional and interactional expressions of the social bonds among the members is closely related to the economic activities of the principal breadwinner.

The poorest households in Camurim still cook on adobe brick

stoves but charcoal is becoming more expensive and many are switching to butane gas stoves. The kitchen has a triangular storage rack for cooking utensils and a cupboard for plates, cups, saucers, bowls, eating utensils, and food. The cupboard is an important status indicator. Its presence shows that the owner sustains his family so well that he needs a place to store food (Heye 1980:120). Increasing numbers of households are buying refrigerators to keep food. The pantry contains cleaning products, brooms, workclothes, and large metal basins for bathing and washing. This space may be converted into a bathroom with a cold water shower and a toilet.

The bedrooms are strictly for sleeping. Children who play on the beds or, even worse, eat in bed are severely reprimanded because they might fall ill. Only the sick eat and rest in bed during the day. The front room is also an improper place to eat. The family only eats in the front room on special occasions such as mourning a deceased member or the celebration of a baptism, Holy Communion, wedding, or birthday. This sharing of food serves to reinforce the social relationships between the nuclear family and the community and therefore belongs in the front room. Routinely, the nuclear family eats in the intimacy of the kitchen, safely away from the street with its potentially harmful influences.

The bedroom is the place of physical recuperation and sexual reproduction. In this sense, it resembles the domestic world as seen from the outside. From the perspective of the community, the house is a place to which men retire from their economic and public status to reproduce the basic social unit of Camurim. Sharing the conjugal bed is the most crucial aspect of the marital relationship, the ultimate obligation of husband and wife. A fisherman may beat his wife and children, neglect their welfare, and fail to buy them food, but the community only considers the household as dissolved when husband and wife no long sleep together. Yet, how that union is worked out from day to day reflects the importance given to the domestic world in the lives of the couple. The prolonged absence at sea of boat fishermen and the daily presence at home of canoe fishermen indicate a different measure of interests.

The parents occupy the bedroom closest to the street. They sleep in a double bed with the foot always turned away from the street and preferably away from the room's doorway. A bed pointing toward the street is associated with death because a deceased person leaves the house feet first. Since the bedroom is associated with life, with procreation and reproduction, a sleeping position that resembles a corpse might effectuate an untimely death. The parents' room has one chair and a trunk under the bed, in which documents, money, and other valuables are stowed away. Clothes, sheets, and blankets are kept in a small wardrobe or chest of drawers.

The children's bedroom is located in the heart of the house, protected by the kitchen and the parental bedroom from the harmful influences of the street. The beds in the children's room point in the same direction as the conjugal bed. The room has a chair and a few boxes with the children's belongings. The parental bedroom has a window in most houses, but the children's bedroom is usually windowless. Windows make the sleeping place vulnerable to undesirable intrusion and undermine parental control over the children at night. When the children reach puberty, boys will invariably sleep in the front room and girls will remain in the bedroom. Parents are much more concerned about their daughters' chastity than about the nocturnal escapades of their sons.

The social context of the front room resembles that of the public domain in that people manipulate social relationships with community members. An important distinction between the front room and a bar is that, in the house, the fisherman is in complete control of the social context. The canoe fisherman Casimiro decides what to convey about himself and whom to admit to his home. He exercises this right even in relation to economic activities in the home.

> Iraçema, like so many other women in households with refrigerators, sells homemade popsicles (*geladinhas*). The customer leans on the bottom half of the Dutch door and asks for a particular flavor. Several times, however, I saw Casimiro turn away a few teenagers. "I don't like these kids," he said.

> "But if you don't want to sell to those kids," I countered, "then why don't you tell your fish peddler to refuse fish to customers you don't like either?" "That's different. These shameless kids (*sem-vergonhas*) have to enter my house, while fish is sold in the street."

In this situation, the anonymous relationship of buyer and seller became personalized by the domestic context in which the economic transaction was to take place. The economic domain threatened to invade the household but was barred by giving a greater importance to harmony and control of the domestic world than to financial gain.

The front room in the homes of most canoe fishermen and boat fishermen contains a formica-covered table with four chairs, a couch and two armchairs with vinyl upholstery, and a cabinet in which some cheap ceramic ornaments and the best dinner plates are displayed. The poorest households do not have these luxuries. Wooden tables, chairs, and shelves take the place of more expensive furniture.

The decorations on the wall of the front room represent the three domains of the social universe of the fishermen: work, family, and community. Without exception, all houses have a calendar, a pictorial representation of Jesus Christ, Mary, or the Holy Family, and a large drawing of the members of the nuclear family. The calendar serves to indicate the lunar cycle, days of the year, and religious holidays. Information about the moon and the tides is, of course, indispensable for the fishermen. The religious images relate the family to a congregation of Christians and show their belief in a Supreme Being, and a church. The family portrait, finally, depicts the affinal, consanguineal, and emotional ties of its members.

Although the sociospatial divisions and categories described above are part of the domestic habitus in Camurim and therefore invite their substantiation through social interaction, they do not determine domestic practice. The spatial structure itself is hegemonic but its appropriation and social definition in domestic practice are not. How the room and the objects in it are used demonstrates the relation between the house and the community. Boat fishermen do not use the front room as a place to meet with friends and, therefore, do not purchase luxury items with the same intention as do canoe fishermen. As will become clear in chapter 7, the economic and public aspirations of the male head of the household influence the importance he will give to his family, how he will interepret the sociospatial habitus in domestic practice, and which links will be drawn between the divisions of the house and the domains of society.

## Architectural Symbols

The porch and the façade are the two principal exterior architectural features that distinguish the houses of Camurim. The porch adds another gradation to the transition between the house and the street. It is an intermediate space between the front room and the sidewalk (DaMatta 1981:71). Only the houses of boat owners and rich townspeople have porches and chest-high walls which surround their property. When Rubens, the town's handyman, wants to enter a house with a front porch, he waits outside the gate, claps his hands, and asks for permission to approach. Because there is a large status difference with the house owner, he will conduct the conversation at the porch. Rubens will again ask for permission to enter and will often leave his thongs outside when he has to step inside. A house that stands independent of the neighboring residences is a sign of high status, not only because of the obvious cost involved but because such demarcation of space tends to isolate the household from public scrutiny. The houses of fishermen, on the other hand, are divided by thin walls and fences. Neighbors can overhear each other's conversations, smell the food that is being prepared, and peer into each other's backyards with little physical obstruction. Parents, therefore, discipline disobedient children with a raised voice to demonstrate their high moral standards to the neighbors. Gossip, instead, is exchanged in hushed voices.

Only about half of the boat owners have open front porches, but others will certainly construct them in the future. The porch is a distinctive feature of the houses of the local elite and, in their desire to belong to that group, boat owners emulate the architectural cues of status. The front porch constructed, the next improvement is a room near the kitchen for a live-in maid. All households of boat owners have servants, but only two have adequate sleeping facilities for live-in housemaids.

The second architectural distinction of Camurim's houses is a particular kind of façade, locally called a *pratibanda* (*platibanda*). A pratibanda is a façade which rises above the gutter and preferably makes the tile roof invisible from the street.[17] Each façade has a uniquely designed top of turrets, multicolored steplike ornaments, bas-reliefs, and sometimes an emblem of a favorite soccer team from Rio de Janeiro. The houses of boat owners, canoe owners, and well-off canoe fishermen and boat fishermen all have pratibandas. Low-producing fishermen and many canoe fishermen generally live in houses without the distinctive façades. The façade is constructed as a false front to the wooden structure and only later linked to the house—when the owner can afford brick walls. For boat owners and boat fishermen, the façade is just a wealth indicator, but for canoe owners and canoe fishermen it has a strong symbolic value.[18] The façade symbolizes the relationship between the household and the community and indicates the place of the owner in Camurim's social hierarchy. No two pratibandas in Camurim are the same in color, shape, or size. The façade gives individuality to the house and identifies the inhabitants through its unique appearance. The fourth-house-from-the-left has suddenly become the yellow-house-with-the-pink-merlons.

In a region dominated by planters and ranchers, canoe fishermen want to make clear that their homesteads are not to be mistaken for the identical cottages of plantation workers. The pratibanda implies that the inhabitants own the house, cannot be expelled, and can deny anyone entrance. The house cannot be violated by what comes from the outside, from the street, or from the powerful who dominate the public world. This autonomy symbolized by the unique façade further implies that the owner is not the serf of a plantation owner but is an independent producer in charge of his household.

## *The House as an Expression of Hierarchy*

The social hierarchy of canoe owners and canoe fishermen is based on the quality of their domestic life. Capital ownership and knowledge of the fishing grounds give respect, and one's economic performance is, of course, also important because it sets the parameters of consumption, but good revenues give prestige in the peer group only if they are spent on the household. Buying drinks for others raises disapproval about the money gone to waste at the expense of the family. The façade serves as a first indicator of a household's standard of living and conveys the fisherman's intention of satisfying the needs of family members. The pratibanda is the only exterior show of wealth because all other signs are kept in the house.[19] The principal reason that luxury goods are bought is to satisfy domestic needs. There is, of course, an important element of conspicuous consumption involved, but the use of these luxury items is always turned first toward the family and only secondarily towards the community, which is quite different from the public display of wealth in bars by boat owners and boat fishermen.[20]

Domestic luxury goods allow their owner to dominate and control social relationships in the front room through his objects. Casimiro has the right to define the terms of social interaction just as he guides the conversation at netmending parties. He dictates the context of interaction, whether those present will watch television, listen to the radio, play dominoes, or talk about politics. He makes his possessions accessible to others, who unite with the household members in their use. The recreational enjoyment bonds those present and enhances the prestige of Casimiro, whose hospitality in temporarily sharing his wealth caused this atmosphere to arise.

The front room reaches its fullest social definition in the house of canoe owners. Casimiro shares luxuries with the less fortunate of his corporate group and neighborhood, just as he makes the largest contribution of fish for redistribution to the turmas. He is treated with respect, is rarely subjected to joking behavior, and his opinions on politics, economic and moral issues are taken seriously. Many canoe fishermen have not yet reached the level of wealth common in the houses of canoe owners but they pursue the same goal and try to follow the example of their higher rank colleagues.

Despite all these manifestations of conspicuous consumption, prestige and social rank are mainly fortunate corollaries of the domestic standard of living; they are not sought after to improve the economic status or gain political power and public influence. The attitude of canoe owners who do not belong to a corporate group shows this clearly. Natalício closes the front door in the evening to all but a few relatives. He thoroughly dislikes comment from the "tongue of the people" (*a língua do povo*) on his material well-being as an improper intrusion in his private life. For the nonaligned canoe owners, the front room is hardly a place for prestige-bearing consumption.

Many luxury goods in the houses of canoe fishermen and boat fishermen are very similar in design, color, and material. The goods are often purchased on installment from large warehouses established in Serrania and other cities inland. The trucks which visit Camurim once a month are overloaded with beds, refrigerators, television sets, radios, bicycles, and furniture. The stores in Camurim do not carry these goods. They do not sell large enough quantities locally to purchase the wares at low prices, while the saline sea air makes the metal objects rust within months. Trucks deliver their goods to many towns and villages in the south of Bahia, all carrying the same brands. This homogenization of manufactured goods facilitates comparisons in material wealth among the households.

Luxury goods in Camurim may be similar in all households but they do not objectivate the same practices for their owners, have not been bought with the same intentions, and do not mediate the same forms of interaction. A canoe owner buys these goods to make his house a comfortable place to live in, for himself and his wife and children. He does not seek entertainment in public places and therefore values a radio, a television set, and an armchair which allow him to relax at home. The house, the pratibanda, the front room and its luxury goods objectivate the canoe owner's interest in the domestic world.

The quality of the standard of living elevates all family members to a higher social rank. The wife plays an important role in prodding her husband to work harder and buy goods that indicate the family's social ascendence. She comments that the neighbors have a new tablecloth, bought a blender, or are planning to reconstruct the pratibanda. Such information may be mentioned casually but it is well-aimed at the husband's pride and at his responsibility of attaining the highest status that his income can afford. Recent purchases by others threaten to diminish his rank and motivate him to work harder. This ranking system is of significance only to the group of canoe fishing households. Outsiders may recognize differences in wealth but they regard canoe owners and canoe fishermen as one undifferentiated group of lower class fishermen.

The possession of luxury goods is an important measure of a household's living standard but does not necessarily guarantee the well-being of the family, which is the condition that finally determines social rank. Clothes and food are very important ways to demonstrate the current living standard. Furniture and a refrigerator could have been bought during a lucky fishing season, but the regular purchase of new clothes for the entire nuclear family shows that the relative prosperity continues. People know the price of clothes from vendors at the Sunday market and hawkers who go from door to door. Food, and in particular meat, is also a considerable prestige-bringing commodity. The consumption of any foodstuff outside the usual diet of rice, beans, bread, manioc flour, and fish sets the household apart from others. These products are bought unobtrusively but meat, and to a lesser extent poultry, are displayed openly. Most food purchases are made by the wife or her daughters, but meat is bought by the husband.

> "I feel like eating meat today," began Silvestre as he approached the three fishermen standing at the streetcorner near his home. "When you're at sea all day, and mess around with fish, you want to eat something different at night . . . Eh, kid!" he winked at his fish peddler. "Here's six hundred cruzeiros to buy meat at Roxinho's." Silvestre would have gone himself if he would not have felt too tired

to stand in line. The boy returned after twenty minutes with a bulky package, and instead of delivering it to Esmeralda, he gave it to Silvestre. Silvestre leaned on the windowsill and unwrapped the large piece of meat from a bloodstained sheet of gray paper. "A good piece of meat," he said, while weighing the chunk in his left hand. "It's getting expensive to buy meat, but it gives me strength, and it's good for Esmeralda and the children."

Fishermen who do not try to improve the living standard are regarded as negligent, uncaring fathers and husbands. This is obviously a self-serving exaggeration but the desire for honor and prestige, and the social pressure from colleagues and family members, help to channel their earnings into the household. However, hard work to satisfy the consumption demands of the household is not just motivated by the desire for social prestige but arises from a genuine caring for the nuclear family. The canoe fisherman relishes the prestige that accompanies the standard of living because of an emotional attachment to his family, an attachment which grows from his continued presence at home. Social prestige is more a spinoff from the domestic quality and standard of living than its goal, because a canoe fisherman's public reputation comes from receiving the community into his well-provisioned home rather than by competing for it outside. Unlike boat owners and boat fishermen who spend much time at sea, he is aware of the many demands of the household and therefore develops a greater compassion for his family. The façade, the front room, and the appearance of the inhabitants are the foci of domestic consumption, because these features best communicate his intentions from the house to the street.

# 7
# Social Relationships and Family Life

The assault was talked about for a long time, and even reached the national news media. The castrated man only ventured out of his house armed with a revolver, ready to shoot anybody who made fun of him. The women in town approved of the wife's actions, "She did right. She must have been desperate and could no longer take his abuse. This will tame (*amansar*) that animal." The men were less appreciative, "She did right to defend herself but she shouldn't have branded him for life. Instead, she should have left him. Who will ever want to marry such a woman?"

The middle-aged man had a history of battery and abuse. His young wife was often marked with bruises and cuts, and the neighbors could hear her screams at night. On the day of the incident, he had beaten her again and then ordered her to prepare some hot water and give him a bath. She boiled a kettle of water, poured it in a basin, and approached her husband. When she got close, she suddenly threw the pail of hot water at him and burned him terribly. He got into a rage and began to beat her as he was restraining her between his legs. In utter desperation, she reached up, bit into his scrotum, and wrested off a testicle. As he screamed in pain, she tried to tear off the remaining gland. Then he fell into a coma. He was taken to the hospital and she left for Vitória.

This revenge for conjugal mistreatment is rare in Camurim and in Brazil as a nation. Battery and sexual assault of women by their husbands are common abuses that seldom reach beyond the walls of the home. Yet, when a man is wounded in his pride and masculinity by a woman who does not take the beating in passive silence, then something ruptures in the practice of male-female relations. The incident sets off a discussion among men about the proper way to handle women, and compels them to cast out the male who allowed his wife to get the upper hand. Women, in turn, are reminded of their routine debasement by their husbands in the home, and the superior attitude of men in public and economic affairs that have pervasive consequences on the household. Relations are reexamined, statuses are redefined, and variations in the conduct of different groups of men come to light in the scrutiny that propels the discourse.

Boat owners and boat fishermen have a narrow definition of their status as fathers, husbands, and heads of household. They try to limit domestic consumption to a level adequate for the satisfaction of what they consider to be basic needs. The men believe that household consumption has little or no social implications beyond the home and should therefore be kept to a minimum so that as much income as possible can be diverted to the public world. The major impetus to raise the standard of living does not come from the boat owner or boat fisherman but from his wife who demands a lifestyle compatible with her husband's economic and public status.

Most women in Camurim do not occupy central positions in the economy or the public arena. They can only represent themselves to the community through the relational status of their husbands. The house, therefore, takes on a symbolic meaning for women which is not acknowledged by boat owners and boat fishermen. Matrimony, the nuclear family, and the establishment of an independent household may have meant more to the men at the time of their wedding but its significance has receded into the background of daily routine, while other more pressing concerns and interests have gained importance. The quality of domestic relations is affected by their emphasis on obligation and transaction rather than consideration and beneficence.

Canoe owners and canoe fishermen interpret the house as a cultural focus which conveys—at the same time—an interpretation of the domestic and the economic world. The interconnectedness of canoe fishing and the household entails an overlap of obligations. The responsibilities of a canoe fisherman at sea are related to his responsibilities at home. A poor economic performance harms the household, and distance from the nuclear family—whether emotional, or in time and activities shared—weakens his motivation to work hard.

Boat fishermen and canoe fishermen may seem to share the same norms of behavior and, in principle, agree on their obligations toward the nuclear family, but their actual practice varies considerably. Where one interprets domestic practice as an expression of unity and autonomy, the other reads only utility. Due to the intimate association of canoe fishermen and their families, the house, its rooms and belongings, the family members, their relations and obligations, and the quality of life, lifestyle, and degree of domestic harmony—all contribute to the maintenance of the house as a cultural focus, while the distance between boat fishermen and their wives and children lessens their involvement in the household.[1]

In the previous chapter, I have described the material and spatial habitus of the division of Camurim society into three domains, and sketched its correspondence with the division of the house into three areas. In this chapter, I will pursue the social dimension of this relationship. The social interaction in the kitchen, bedroom, and front room resembles that of the economic, domestic, and public domains. The nuclear family, the basic unit of Camurim society, is reproduced in the kitchen, conceived in the bedroom, and interacts with the community in the front room. In a comparable way, society is reproduced through the economy, conceived in the home, and manipulated in the public sphere. The correspondence of these two levels of social complexity—house and society—is the result of an interpretive process in which social practice is understood in part from the perspective of the domestic world, a perspective acquired during early childhood and perpetuated throughout life. Domestic practice provides a conception of the quality of social relationships which then serves outside the home. The transactional quality of certain domestic relationships can be found in a similar though transformed way in society at large. For instance, the material obligations of a man to his family correspond to his contractual obligations at work, the conspicuous consumption in the front room resembles the competition for status in public, and the importance of sexual loyalty in the bedroom is related to the significance of the domestic world in the lives of the couple. These influences do not go in only one direction. The relationships among the social domains of society influence the interpretation of the sociospatial habitus of the domestic world.[2]

I will demonstrate that the domestic habitus in Camurim is more dynamic, and far less autonomous and hegemonic than Bourdieu's definition of the term suggests. Contemporary domestic practice in Brazil is set in a spatial habitus imposed on colonial society in the eighteenth century, yet family life inside is mutually dependent on social activities outside the home. The content and quality of the inhabitants' social statuses and relationships in the economic and public domain greatly influence, and are influenced by, their interpretation of the domestic spatial habitus. Even though the spatial structure itself is hegemonic, its interpretation and social appropriation in domestic practice are not.

I will first discuss the concepts used by Camurimenses to describe social relationships. Each social domain is characterized by a particular type of interaction. This conceptualization is unambiguous when defined discursively, but it leads to considerable interpretational controversy when used to categorize social relationships according to their intent and meaning. Actors may place their statuses in other referential frames of rights, duties, privileges, favors, and debts, which make one person consider an obligation what the other regards as a gesture of friendship. This analysis of interpersonal relationships will be followed by a description of the formal obligations of married couples and how social norms are modified in practice. The social interaction in the three significant social spaces of the house demonstrate the difference between the house as a cutural focus and the house as a bounded domestic domain. This divergent appreciation works its way into the social relationships outside the home and, in turn, reflects back on the domestic world leading to differences in the way house and society become related.

A brief discussion of household budgets shows the difference in the fulfillment of conjugal obligations among the various social groups, since the social position of a woman in the community depends on her husband's interpretation of his conjugal obligations. Finally, I will elaborate on the interpretation of the house as a cultural focus among canoe fishermen and analyze the reasons for its more peripheral significance among boat owners and boat fishermen.

## *The Quality of Social Relationships*

Zé Peroba enjoyed his new status. He had married the day before and acted with self-confidence amidst the men gambling in Chandú's bar. "You're macho now, eh?" joked Caetano, the captain, as the group began the game. Each participant tried to guess the total number of matches they held hidden in their fists. After a dozen rounds, the loser would buy a bottle of beer. Fifteen minutes after the game got under way, Zé Peroba's wife, Rosa, appeared at the other side of the street trying to get her husband's attention. A seven-month pregnancy glowed in her face and she seemed genuinely happy that Zé had married her. Rosa gestured with her hand but Zé ignored her. "Peroba, your wife is already calling you. She must want some more," chuckled Caetano. "Exactly," continued another, "once women get the taste, they always want more." Zé Peroba joined the men in their laughter but, as he glanced askance at Rosa, said softly but loud enough for the men to hear, "Vai tomar no cú!" ("Don't bother me!").[3] He wanted to convey to his friends that he had complete control over his wife and that he did not allow her to meddle in his activities. Rosa finally left. Her smile was gone and the expression of disappointment made me think about the many expectations she must have had before the wedding and the many disillusions that were still ahead of her.

Men and women in Camurim acknowledge the importance of love in any intimate relationship, but they also believe that romantic love confessed during courtship must be proven in marriage through acts of consideration and the fulfillment of obligations. Obligation (*obrigação*) refers to the expectations of two persons with respect to their mutual rights and duties. "Obligation," explained Geraldo, "is to do things even against your will. You *have* to fulfill an obligation." A failure to perform those expected duties, and disagreements about the reciprocal obligations will temporarily disrupt the social routine and make the actors reconsider their relationship.

Consideration (*consideração*) refers to the anticipation of another person's needs without, however, experiencing their fulfillment as a duty or obligation. Acts of consideration are not direct responses to expectations but emerge from an emotional attachment to the other. A person who acts out of consideration does not experience duties as obligatory acts but as voluntary demonstrations of affection, esteem, and respect. Although they seem to be close in meaning, consideration and love (*amor*) are not synonyms. Love is the emotional and affective attraction of a man and a woman.[4] Consideration is the continuous demonstration of that love in deed, gesture, and intention.

> Juliana prepares the evening dinner and washes the family's clothes out of consideration, not because she thinks it is her duty. She used to get up at three o'clock in the morning to prepare polenta with coconut milk which a teenage vender would peddle in the street, just to have more money to buy her children some nicer clothes. Edinho bought her a stove to relieve her work, and now he is planning to buy a refrigerator. Juliana knew about this and she showed me the shirt she was making to surprise him the day the refrigerator arrived. These tokens of affection and generosity are imbued with true feelings of caring that heighten their mutual consideration.

The conjugal relationship, like all social relationships, goes through emotional ups and downs. Obligation and consideration describe the long-term standing of social relationships. *Média* is a short-term indicator, an emotional barometer of the understanding between two persons at a certain moment. *Média* evaluates the everyday mood and temper of actors in a relationship which is already based on obligation and consideration.

> When Juliana discovered Edinho's visit to a brothel, she was seething with anger. She went to the kitchen and slammed the skillet forcefully on the stove as she fried some fish, Edinho and Juliana recalled laughingly as they related their quarrel to me. Her anger did not diminish her consideration for Edinho but she performed the domestic tasks with less dedication than before this incident. Edinho, however, soon forgot the episode and continued to spend his money on prostitutes and liquor, and he neglected his resonsibilities as the head of their household. The continued low *média* reduced Juliana's consideration until there was only obligation left. She had so much resentment for Edinho that she began to deny him sexual intercourse and prepared the food to her taste, not the way he liked it. Now, Edinho began to complain and his consideration for Juliana dropped to a point that he only performed the most basic duties in the household.

Spouses who have lost their mutual consideration might still live up to basic domestic obligations but the conjugal relationship will become very transactional and calculated. Juliana continued to wash and cook, and Edinho kept buying food and paying the utility bills but they did so out of a sense of duty, not with affection. Yet, they remained together as long as there existed at least a satisfactory flow of transactions among them. Furthermore, the persistence of a household in strife still allows for the opportunity of improved interaction because the *média* does not remain low at all times but will fluctuate between occasional highs and extended lows. A prolonged high *média* might give rise to renewed consideration since conjugal estrangement does not necessarily destroy all feelings of love. People can quickly break and replace economic and public relationships, but they endure considerable emotional strain and long periods of conflict before they decide to disengage domestic relationships.

Each social domain is characterized by a distinct yet ambiguous combination of obligation, consideration, and *média*. Obligation predominates in the economic domain. Consideration is the prevalent but not exclusive quality of domestic relations, and *média* characterizes the public sphere. Although the quality of the social interaction in each social domain can be generalized in this way, the contextual ambiguity of the relations is a major source of contestation. There may be considerable disagreement between two persons about which terms describe their relationship best. Power differences are played out through this ambiguity, allowing one person or group to manipulate the reciprocal relation to its advantage.

Economic relationships are mostly based on contractual obligations. People have to perform the tasks assigned to them in exchange for a share of the catch, and nobody expects them to work with consideration. "At sea," said Natalício, "you even have to help your enemy when he is in danger." The economic relationship is contractual and not supposed to be affective. Still, persons who have different economic positions may be in conflict on how to classify their relationship and the actions that substantiate it. Boat owners claim that they allow boat fishermen to take fish home out of consideration. "I do not have to part with this quantity of fish," said one, "but I feel compassion for their families." They emphasize that this custom should not be taken for granted and should instead be interpreted as a favor, because formal obligations, not privileges, are the basis of economic relationships. However, a boat fisherman argued, "It is my right to take fish and he has to give it." Similar conflicts arise over the *vale* and the social assistance to the families of crew members. Boat owners regard these services as favors, while crew members believe it to be the obligation of capital owners to extend credit and provide help when boats are at sea.

The interpretation of an action as either obligation or consideration is important. When an act of consideration can be reclassified as one of obligation, then the recipient no longer has to reciprocate with equally valuable favors. The act becomes part of the rights and duties of their relationship. Despite these interpretational conflicts, most interaction between boat owners and boat fishermen is based on obligations, even though many hours spent together under life-threatening circumstances or a good personal relationship ashore and a high *média* may occasionally make room for gestures of consideration. However, when the *média* drops, these favors cease immediately and only the necessary tasks will be performed.

Economic relationships among canoe fishermen are also contractual but they are more complex because of the existence of corporate groups and the affinal, consanguineal or affective closeness of fishing partners. In chapter 3, I explained how corporate groups redistribute fish during hard times and how canoe fishermen will change the catch division to meet each other's needs. Their economic relationships are more often imbued with consideration than those among boat owners and boat fishermen. Friendships develop readily and add consideration to the economic relationship. Yet, this consideration is always measured with the yardstick of obligation. The close companionship between two canoe fishermen will inevitably dissolve if one does not carry out his tasks properly.

The concept of *média* prevails as the principal characteristic of public relationships. *Média* is not only a short-term indicator of dyadic relations but it may in itself also be a unique quality of reciprocal relations. People who have recently made each other's acquaintance may develop a *média*. Public relations are loose, predominantly voluntary alliances. Two persons must first initiate a mode of uncommitted communication before they can "establish a *média*" (*fazer uma média*). They must have each other's acquaintance (*conhecimento*) by recognizing each other in the street and exchanging formal greetings. Eventually, this relation may evolve into a more permanent form reinforced with conversations about each other's work and political sympathies. Isidoro wants to buy a boat and tries to persuade a planter to be his cosigner. The two men exchange greetings, inquire after each other's health, and have drinks together. They are on friendly terms and Isidoro boasts that he has "a good *média*" (*uma média boa*) with Sr. Barbosa.

A persistently high *média* may develop into a close friendship with mutual consideration. The consideration will find its expression in more tangible exchanges such as loans, information about fishing grounds, gifts of fruit, assistance in quarrels, and sometimes introductions to each other's social contacts. After a certain period, these gestures of consideration tend to become obligations. Friends are supposed to be able to count on each other and respond to certain expectations.

Canoe fishermen accuse boat owners and boat fishermen of insincere manipulation, and of disguising their personal interests as friendship. Celestino gave his opinion on "the practice of establishing a *média* in the street" (*a prática de fazer uma média na rua*), "Nobody has a *média* to give or receive help. This [street] *média* exists only in the exchange of rum, and hence there is not going to be any consideration." Boat owners and boat fishermen deny that their friendship is feigned but admit that public relationships are expected to expand beyond informal niceties. The public domain consists mainly of communication networks which have tangible consequences in all domains of society. A person with a good network can get educational and medical assistance for his family, loans to buy capital goods, and preferential treatment from the municipal authorities.

Obligation, consideration, and *média* do not have the same importance for all men working in the fishing economy because they are not equally involved in contractual, voluntary, and affective relationships. Canoe fishermen do not spend much time enlarging their circle of acquaintances, while boat owners and boat fishermen are often not at home. The relatively little exposure to voluntary relationships in which manipulation is paramount reduces this quality in economic and domestic relationships and diminishes its importance in the limited public contacts of canoe fishermen. In turn, frequent absence from home and the ensuing emotional distance seem to make boat owners and boat fishermen more prone to manipulation and personal gain, and tend to diminish affection and consideration in their social relationships.

This differentiation among boat fishermen and canoe fishermen becomes most apparent in the home. Domestic relations are multistranded, a complexity which is objectivated in the spatial habitus of the house. The domestic relations are imbued with consideration, but each social division in the house adds a different combination of obligation and *média*. Ideally, consideration prevails in the bedroom, *média* in the front room, and obligations are most important in the kitchen/veranda area. The following section will describe how boat owners, boat fishermen, and canoe fishermen express the domestic relations differently with respect to the three qualities of social relationships analyzed above.

### Household Relations and Domestic Obligations

The men of Camurim consider the obligations of husband and wife to be complementary but imbalanced. They agree that without their material support the household would not exist, while domestic services can always be done by servants. Women think the opposite as Dona Consuela, a seventy-year old woman who has had twenty-six children of which only four survived, told me one day when her son-in-law arrived drunk at her house.

> "A woman is superior to a man because she has more strength. A man doesn't dominate a woman, but a woman dominates even a fierce man." "How?" I asked. "A man needs a woman to wash clothes, to give him food, and those type of things." "But can a man not pay a woman to do these things?" I asked. "And if he is poor?" Dona Consuela replied. "If he doesn't have money to pay somebody to do all this work for him? Ah, a woman has more strength. There are women who live their entire life alone, but a man cannot live alone. Have you ever seen a man who lived alone?"

In an unusually frank exchange between Fábio, the canoe owner, and his wife Márcia, the different understanding of the conjugal relationship by men and women was expressed in the following straightforward manner.

> "A man marries a woman," explained Fábio as we were sitting under the veranda of his house, "so that she can re-

> solve his problems in the house, such as cleaning, washing, and taking care of the children." Márcia interjected with indignation, "A man marries to beat and mistreat his wife!" Fábio raised his voice, "A woman marries because she likes to and to have respect. She earns respect in the community by being a married woman, a mother, and the head of the house (*chefe da casa*). For this reason, a man is superior to a woman and therefore he has the right to go to the street [i.e. commit adultery]." "And the woman does not have any rights?" I asked. "But if it is the man who gives her her rights?" Fábio replied rhetorically. "A man has more rights to have another woman, because he walks a lot in the street because of his work. A woman does not, because she stays at home. A woman cannot do this. If not, she loses her rights to her husband. Only her man has the right to her, for this reason there exists marriage."

Men, therefore, try to resort to common-law unions (*amasiado*) or, as they say, "to marry behind the door" (*casar atrás da porta*) or "to leave the door open" (*deixar a porta aberta*). Both expressions refer to the status transformation people undergo when they move between the house and the street, between the intimate, enclosed domestic domain with its mutual obligations and affective relationships, and the open public world where men can engage in voluntary, short affairs with any willing partner. The first expression indicates that the common-law union is not sanctioned in public but consumed and acted out in the house as if the couple was married. The second expression implies that men can dissolve the cohabitation arrangement and reappear as uncommitted men if they are unhappy with the reciprocated services (cf. Azevedo 1965:296–298; Ribeiro 1945). In a variation on the same theme, Raimundo said, "When I leave the house, I'm a bachelor. I can do whatever I want without my woman saying anything about it."

Yet, the infidelity is not mutual. A woman can, of course, never assume such an ambivalent attitude toward men. She would quickly be called a "street woman" (*mulher da rua*) or a "loose woman" (*rapariga*) who brings disgrace on her husband, children, and family. This double standard of morality which protects the purity of women and gives sexual freedom to the men has been attributed to the instability and moral decay of the early colonial days (Azevedo 1965:290), but can be traced further back to ancient Greece (Foucault 1986:145).

However, women are increasingly more reluctant to enter amasiado relationships. One or two decades ago, they did not want to lose their virginity without any formal commitment and they were especially afraid to fall to the low status of a rapariga after a partner abandoned them and their children. Unable to marry as single parents, they could expect only unstable affairs with men and might eventually feel forced to become prostitutes in a community with little steady employment for lower class women (Kottak 1977). This concern still plays some part but teenagers have become sexually less inhibited and society more tolerant. Young lovers are no longer dragged to court by the girl's father to enforce a marriage, and men seem less resolved to marry a virgin.

Overall increased material wealth has made women more aware of the importance of a civil marriage in case of divorce. Concubinage does not give a woman the legal right to claim alimony and a share of the property. Separation of common law unions results in the woman and her children returning to her parents, while among married couples the husband generally leaves the house. Civil marriages enhance the woman's negotiating power whenever conflicts arise. Before divorce was legally recognized in Brazil, men preferred religious to civil marriages. Religious marriages lacked any legal protection for the woman but, at least, did away with moral objections to concubinage. The local Roman Catholic priest, however, is following the directions of the Brazilian ecclesiastical authorities and is demanding that civil weddings must precede all religious ceremonies. The insistence on civil marriage has, therefore, reduced the power of men in dominating women with the threat of abandonment.

A man and a woman enter marriage with a conception of the conjugal relationship acquired through socialization during childhood and adolescence. On the whole, the husband expects his wife to bear children, prepare meals, wash and sew clothes, clean the house, look after the children, be loyal, and respond promptly to his sexual desires. The wife expects her husband to provide shelter for the nuclear family, show sexual interest in her, and pay all household expenses such as food, rent, utility bills, school fees, clothing and medical costs. With the exception of low-producing boat fishermen, most men and women—by and large—try to live up to these expectations, but the way in which the duties are performed differs from one social group to another. A close look at the social interaction in the front room, kitchen and bedroom, the correspondence between these domestic areas and the three social domains, and the different manifestations of consideration and *média*, reveal the divergent interpretation of formal obligations among the different social groups.

> Silently, Esmeralda observed her husband and his canoe fishing colleagues as she sat away from the front door and close to the kitchen. She attended the men quickly when Silvestre asked her to prepare coffee and some snacks while they were playing a game of dominoes. Esmeralda never invited her friends in the evening, but when her husband was out, she usually chatted to neighbors and relatives in the kitchen, avoiding the front room as a predominantly male space.

Wives of boat owners and high-producing fishermen have taken over the front room as their public space. The boat owner, Arnildo, buys luxury goods and employs domestic help more as tokens of conjugal responsibility and out of an obligation to give the household respectability than as ways of expressing his consideration for the nuclear family. The wives of boat owners are not burdened with domestic chores because servants do the work for a few meals a day and a monthly salary of Cr$3,000–5,000 ($6–10). Solange can, therefore, maintain an active public life of social calls to the wives of other boat owners and members of the local elite. She challenges Arnildo to buy the latest luxury items available in Camurim and Serrania in order not to fall behind the other women. There exists a danger of conjugal conflict because Arnildo wants to spend part of his earnings in bars and brothels, but he fortunately earns enough to satisfy both demands.

The dilemma is more difficult to solve for high-producing fishermen such as Chico Cazuza. His wife is often too busy to use the front room for visits during the day, but at night she likes to leave the house to watch soap operas (*novelas*) or receive other women at home to show off a recent acquisition.

> "Jumelice just wants me to buy new things all the time," lamented Chico Cazuza. "First, she complained that she couldn't cook on wood anymore. I bought her a gas stove and she was the happiest woman in town. Then, our neighbor got a refrigerator. I bought her one. Now, she wants a bathroom." He threw his hands up in the air and looked at me with feigned despair. "You know, Antônio," he began again, "before I got married I had only one name: Francisco Cazuza dos Santos. Now, I have lots of names: Chico Não Tem Café (Chico There's No Coffee), Chico Vou Comprar Pão (Chico I'm Going To Buy Bread), and Chico Quero Uma Televisão (Chico I Want A Television Set). I never hear: Chico Aqui Tem Mil Cruzeiros Para Quebrar o Galho (Chico Here's One Thousand Cruzeiros To Help You)."

High-producing fishermen come close to regarding the household as a contractual liability whose needs must be met from a moral standpoint but whose demands restrict their success in securing the public contacts that they hope will, in the long run, yield far more wealth and prestige than the temporary benefits of conspicuous consumption. Clothing and food are affected most. Luxury items are usually acquired after a fortuitous catch, but the weekly household budget leaves little leeway to buy meat or pay the installments on a piece of clothing. Ambitious boat fishermen want to set aside as much income as possible for their nightly outings and only make those house improvements necessary to avoid their identification with poor fishermen.

Souza and his wife, Dalva, seldom receive people at their home. Their barren house, leaking roof, the dirty façade, and the simple wooden furniture painfully communicate their economic failure. Low-producing boat fishermen achieve prestige neither at home nor in public. They avoid their home when they are ashore and spend their meager income in the small skid row bars near the harbor.

The front room represents the public face of the domestic world for all Camurimenses, but the interpretation in practice shows a great variation. Boat fishermen will, at most, consider the front room as a showcase for their economic success. They seldom receive colleagues at home because they cultivate their social contacts in the street. Luxury goods are often bought on the spur of the moment after a good catch and not primarily out of consideration. Their satisfaction derives more from the public expenditure of a large sum of money, parading the purchase through the streets and showing it off in a bar to their colleagues, then from improving the standard of living at home. Canoe fishermen, as was shown in chapter 6, pride themselves on being able to entertain corporate group members in their front room, and regard the ensuing prestige more as public approval of the high quality and standard of living of their household than as a show of conspicuous consumption.

Many of the obligations between husband, wife, and children come to light in the kitchen. A canoe fisherman takes all his meals at home and is, therefore, confronted daily with the nutritional needs of the household. This continued encounter with the results of his labor, and his wife's many services and tasks, give substance to his obligations and enhance his involvement in the domestic world.

Although Silvestre, the canoe fisherman, pretends to be in charge of the household, the kitchen is as much a female space as the front room is a male space. Esmeralda directs life in the kitchen because almost all her services are done there. Her husband may have the ultimate say on when and what to eat but he rarely exerts this right and she routinely decides on the order of domestic work. In the kitchen, Silvestre and Esmeralda experience the quality of their conjugal relationship and learn of each other's consideration for the family. When Silvestre comes home in the late afternoon, he expects the house to have been swept, the beds made, and the children bathed and fully dressed. Esmeralda will be waiting for him with hot food on the table, fresh clothes, and warm bathing water in the basin. During the meal, he will carefully remove the fishbones from his youngest daughter's plate, and at night he helps his wife shell the large bag of peanuts she hopes to sell the next day. All these activities contribute to the social construction of family life. Even conflict contributes to the maintenance of domestic practice. Quarrels have to be dealt with and a low *média* cannot be ignored, because of negative consequences to the well-being of the family members. The domestic world reproduces itself when the members eat, rest, cook, joke, argue, fight, gossip, and play together. Without this interaction, the household would become an emotionally and socially narrow association of individuals with affinal and blood ties who only act in terms of minimal obligations.

It is exactly this daily intimacy with the household which is lacking among the boat owners and boat fishermen. Arnildo and Chico Cazuza spend relatively little time in the kitchen or on the veranda. If the weather is good, they are only ashore on weekends and, even then, seldom eat at home. They might go on a fish fry along the banks of the Rio Camurim, eat at a brothel, or be invited for a barbecue at the home of a rancher. Domestic conflicts and a concomitant low *média* are rarely dealt with directly because the boat fishermen often leave for sea before the differences can be made up, and return home with the hope that the disputes have been forgotten. I asked Júlio, a captain-owner, why he does not spend more time at home on the few days he is ashore. His answer seems quite insensitive and fails to express his emotional attachment to his wife, but it is representative of the daily interaction between boat fishermen and their wives.

> "I dread staying in the house, I don't like to be tied down in the house. I prefer to leave right away for the street to relax with my friends at the riverfront. I like to talk with some friends and take a beer in a bar on a corner. I do not like to remain in the house much because my wife will ask me to help her, to take care of the pig, of the children. When you spend all your time with your wife, you end up getting sick (*enjoar*) of her face. I hardly spend any time at home. When I go home it is usually only to sleep or to eat,

> but I do not like to hang around the house. People say of a man who stays too much at home that he is stuck under his wife's skirt (. . . *que ele está enrabado abaixo da saia da mulher*)."

The consequences for domestic life is that members of the household have little opportunity to give substance to their domestic obligations and family relations. Jumelice spends more time in the kitchen with the children, but the absence of the father and husband make the family incomplete. The nuclear family exists as a socially recognized unit sanctioned by community and Church but with members whose identification with each other and through each other does not attain the domestic wholeness of the canoe fishermen and their families.

The emotional involvement of both partners in the households of boat owners is even less than in those of high-producing fishermen. Chico Cazuza may be seldom at home, but Jumelice does most of the domestic tasks. Work in Arnildo's household, instead, is done by servants. It is difficult to establish to what degree boat owners' wives, such as Solange, feel attached to their home but it seems as if they regard it in certain respects as a means to enter the public world.

The intention of boat owners to relieve their wives' domestic obligations helps to deprive the women of their status in the household. The work of the servants strips the women of ways to define, justify, and give substance to their roles as mothers and wives. The kitchen and veranda are the domain of servants and children, with—at the margin—a loving but hardly involved mother and a frequently absent father, who sometimes feels resentful when his wife does not fulfill a traditional female role.

> When Arnildo and I walked home around midnight on a warm summer evening in February 1983, he invited me in for dinner. "No, thank you," I replied, "it's already late and I have to get up early tomorrow to watch the beach seines." He laughed, "You and your research, who wants to know anything about those drunkards?" "But Arnildo, tell me," I asked, "how come you're having dinner so late?" "Well," he said, "I took a bath in the river and went straight to the

> bar. I told a kid to tell my wife that I would be coming home late. She will be waiting for me with the food ready." He quickly raised his eyebrows with an improper smile to convey the double meaning of the word "food." I left him at the doorstep and went home. The next morning I found Arnildo with a sullen look on his face. "What's the matter with you?" I asked. "You know what's the matter?" he said angrily. "I die at sea to earn money and make the household run, and my woman doesn't even prepare me a plate of food. She just wants to spend her time gossiping with those friends of hers." He turned in the direction of his house and made an obscene gesture.

In Souza's house, the kitchen is a predominantly female domain. Dalva does all domestic services and earns part of the money as well. Indebted fishermen are socially and economically marginal to the household. They hardly participate in its activities and its costs are met by their wives and the boat owners. Dalva receives the *vale* sometimes from her husband or directly from Raimundo, the boat owner. Her household is incorporated into the production unit because its survival does not depend on the economic effort of Souza but on his membership in a fishing crew. The *vale* will be given irrespective of his performance. When Souza threatened to leave his family without any means of livelihood because he spent the advance on alcohol, the *vale* was given to Dalva. Although thoughts of divorce have crossed her mind, she told me confidentially, Souza still brings fish home and occasionally he tries to do his best at sea and stay away from the bars.

The conjugal relations of boat fishermen are much more contractual than those of canoe fishermen. The economic dimension of domestic life is more pronounced in boat fishing households because of the extraneous demands of the public sphere on limited earnings. Very much as in their economic contracts with the boat owners, boat fishermen negotiate with their wives about household needs that have to be met and will seldom exceed those demands in order not to raise the expectations. The household is a liability which might stand in the way of their career.

Canoe fishermen, instead, infuse social interaction in the kitchen with more consideration because the nuclear family is both their principal source of emotional satisfaction and the principal objective of their work effort at sea. They personally experience the daily economic reproduction of the household and work hard to improve its standard of living.

The bedroom constitutes the third area in the house. Husband and wife expect one another to satisfy their mutual needs for physical recuperation, biological procreation, and sexual intimacy in the bedroom. The conditions of this trust vary substantially between the sexes and from one social group to another. Canoe fishermen have little choice but to spend the evenings and nights at home. The physical exigencies and economic demands of canoe fishing compel them to avoid bars and go to sleep early. Their frequent presence at home makes it difficult for husband and wife to keep illicit sexual affairs from one another. Esmeralda would find it hard to entertain a lover when Silvestre might come home unannounced. Silvestre seems to have more opportunity for sexual liasons during the day, but a lasting affair would not remain unnoticed. Women, whether married or single, who become involved with other men will often demand some material benefits. A wife would quickly know that her husband was giving away fish to another woman or was paying regular visits to a brothel.

> Esmerlda told me about Silvestre's fling with another woman. "Do you think that I would have let that woman take my man? At first, I didn't say anything. I still loved him but I didn't want to show any consideration. I cleaned the house silently, ate silently, and went to bed silently. Later, I told Silvestre right in front of the children to stay away from that woman, and I even went to her house to tell her to back off. Then, I took the kids and went to my parents' house. He really missed them." "That's true," whispered Silvestere. "Well," Esmeralda continued, "I came back home after he promised he would never do this to me again, because I tell you, Sr. Antônio, next time I'll leave for good."

These forms of social control help to discourage extramarital affairs, but the most important reasons for conjugal loyalty are, of course, emotional. Adultery is a betrayal of marital trust and is especially painful because of the symbolism that surrounds the house. The bad influence of the street has invaded and defamed the house, embittering and offending its sanctity and that of its inhabitants.[5]

Boat owners and high-producing fishermen are just as concerned about the sexual loyalty of their wives but for other reasons. They place less emphasis on domestic relations and conjugal obligations but worry about the damage to their public reputation. A canoe fisherman is, of course, not indifferent to public judgment on infidelity but he spends more time indoors and the members of his corporate group will avoid this sensitive topic in his presence. Boat owners and high-producing boat fishermen, however, depend on their public reputation for benefits as diverse as bank loans, honor, respect, and the deference received in public. The failure to prevent one's wife from having an adulterous affair engenders more than shame. It betrays the man's lack of masculinity and implies an inability to properly master other social situations. A man in a male-centered society who cannot dominate a woman is not a "real man" (*homen macho*) and is expected to lose control elsewhere.

Solange has become tolerant although not acquiescent of Arnildo's extramarital affairs. At first she threatened to leave the house but after some confrontations has come to accept the escapades as long as Arnildo fulfills his domestic obligations. As she told me a few days before my departure, "Arnildo stays at home only when he is ill or when he wants to sleep. Since we got married, he has not stayed one Sunday in the house, and he never eats here. If he dies, I am going to miss him because he died and, of course, for the money, but I am not going to miss him in the house." Jumelice, the wife of the boat fisherman Chico Cazuza, is more irritated by the visits to brothels than Solange because of the smaller household budget. Jumelice still has desires for luxury goods and is less involved in extra-domestic networks that could distract her attention from the household. Solange has accepted Arnildo's interpretation that the house and the street are entirely separate domains that should not be mixed up. "Is eating in a restaurant the same as eating at home?" argued Arnildo.

"Well then? A brothel must not be mistaken for the bedroom."[6] The activities performed in these two settings may be the same but the entirely different social contexts give them distinct and incompatible meanings. Seen in this light, it is understandable that a canoe fisherman who is angry at his wife will not come home to eat but will go to a bar or, if he is very upset, to a brothel—acts that would go unnoticed by the wives of boat owners and boat fishermen.[7]

Low-producing fishermen are marginal to all three areas in the house. Women do not feel the obligation to be sexually receptive to husbands who fail to sustain them. Souza has lost the right to consummate his marriage because he does not fulfill the obligations that constitute his conjugal bond. The moral forfeiture of intercourse does not imply a voluntary abstention. Although it is hard to obtain reliable information on this emotionally charged subject, I have heard men telling of being locked out of the house and others who, with a mixed sense of pride, anger, and shame, confessed to marital rape.

A less aggravating way out is a visit to a brothel. A man enters a short-term relationship with a prostitute under clear-cut transactional conditions. Once the negotiated fee is paid and the service performed, the relationship is cut without entailing any obligations or further expectations. Souza does not attend the more expensive brothels frequented by Arnildo and Cazuza but visits older single-parent prostitutes at the edge of town. He does not have enough income to establish steady relationships or pay for the costs of food and drinks in a brothel. His contacts are irregular and restricted to those days when he has made some cash to go on a binge. Instead of at least temporarily assuming his position as the head of the household, he leaves Dalva and the children in destitution.

This irresponsible behavior may persuade Raimundo, the boat owner, to pay the advance to Dalva. The transfer of money frustrates Souza even more. His complete withdrawal lowers the household budget below subsistence and forces Dalva to seek whatever means to raise money. With few job opportunities available for lower class women, an extramarital affair might be a tempting solution. Yet, only very few women are ever driven to this extreme. The majority, like Dalva, survive on small advances and occasional work. Women do not actively solicit adulterous relationships but they are propositioned by boat owners, storekeepers, and rich townsmen when they ask for extra *vales,* buy goods on credit, or seek employment. In exchange for sexual favors, they receive food and some money. The men claim that they do not take advantage of their position of wealth but merely give the women sexual satisfaction while, in addition, helping them in times of economic distress. Adultery completes the process of marginalization of a poor fisherman. Aside from being marginal to the social and economic reproduction of the nuclear family, he also becomes biologically marginal.

> "The man whose woman commits adultery pays his debt twice," explained Dona Consuela. She clenched a knife in her bony fingers as she was cleaning a fish. "You know, Antônio, a poor woman will do anything to keep her children alive. Even if she has to pay with bastards (*pagar com a raca*). He gives her some food and an extra *vale.* When she gets pregnant, her husband doesn't know that the child is not his . . . he is too busy with his bottle of rum. So, he pays with his honor, with bastards, and with his work because he still has to pay his debt. And the poor woman has to suffer all alone, she cannot tell anybody about the pain of carrying the child. This is very sad, very sad indeed."

The affairs are, of course, kept secret but will often become public knowledge after a few weeks or months. The seriousness of this discovery lies not so much in the sexual infidelity of the wife but in the almost irreparable damage to the husband's public reputation. His authority over his wife has been violated (cf. Foucault 1986:146–151), and his masculinity challenged. Seeking public support and retribution, the fisherman will threaten to kill his unfaithful wife and her lover. Revenge is also an attempt to restore his male honor and to avoid being called a "tame cuckold" (*corno manso*) (cf. Willems 1953). He might beat her and verbally abuse the boat owner but he will rarely carry out his threat. Although men believe that adulterous wives should be severely punished, in this case public sympathy is on the woman's side. The low-producing fisherman has neglected the household and his wife's sexual needs. The incident might lead to a temporary separation but not to divorce, if the lover was a married man. All would suffer more by divorce than by a reconciliation based on the promise of improved conduct. The fisherman will try to resume his responsibilities and support the household, even though the affair has left a permanent smudge on his public image.

The penetration of the bedroom by the public domain through repeated marital infidelity reflects the secondary interest of boat owners and boat fishermen in the domestic world as a whole. As I will explain in the next chapter, these men feel more at ease among a group of intimate friends in a local brothel or bar than in the sole company of their wives and children. Canoe fishermen, on the other hand, attribute a great emotional and symbolic value to the house, which considerably restrains the male propensity in Brazilian culture towards sexual adventures and similar shows of masculinity.

## *Income and Domestic Spending*

The seemingly clear-cut sexual division of labor between men and women is complicated by the different ways in which role obligations can be interpreted. Men have to maintain the nuclear family but there are no absolute standards that indicate how they must satisfy these needs. Allocational preferences outside the household reduce the amount available for domestic expenses. Only the collection of statistics on household expenditures can provide the proportion of the consumption items with accuracy. However, these quantitative details are of less concern in this study; what matters are the general differences in consumption of the different groups of fishermen, differences that can be used to delineate their interpretation of domestic obligations.

Food, housing, clothing, health care, and education are the major expense categories in the households of Camurim. Food and acute health care take precedence over other expenses among the canoe fishermen. A canoe fisherman will not allow his family to go hungry or suffer from illness when he has money to buy food and medicine. Such demands, however, always threaten to deplete small savings due to fluctuating income. Canoe fishermen solve this dilemma by tying up income in advance by buying goods on installments. As they are aware of the consequences of a default on the payments, they prefer temporarily to eat less expensive and less varied food rather than lose the item.

The income earned by the wives of canoe owners and canoe fishermen is strictly separated from the household budget. They refuse to pay for domestic expenses because they fear that the men would soon renege on their obligations. During times with very poor catches, some men had to borrow money from their wives to buy bread. A married woman does not regard remunerated work as supplementary income for the household but as entirely beyond her obligations as a wife. She has the right to stop working whenever she wishes, and she can spend her earnings on whatever she desires. The most common source of income for wives of canoe owners is the sale of homemade popsicles (*geladinhas*), while wives of canoe fishermen often work as charwomen.[8]

Although a woman does not pay for basic consumption goods, her earnings still benefit the household. Before anything else, she pays a laundrywoman to wash clothes. Next, she buys clothes for herself, the children, and maybe for her husband as a demonstration of consideration. Any money that is left is usually spent on health care and kitchen appliances. She is, of course, not indifferent to suggestions from her husband. They may sometimes pool their savings to buy a desired luxury item.

As was shown in chapter 5, high-producing boat fishermen earn about the same as canoe owners but their household budgets are at least one-fourth smaller because of high drinking expenses. The household budget is the size of the advance, usually $1\frac{1}{2}$ times the minimum wage. All domestic expenses are paid from this fixed sum managed by the wife. If she complains that the *vale* is too small, then her husband accuses her of mismanagement. After handing over the vale, the high-producing fisherman considers himself released of all monetary conjugal obligations. Therefore, boat fishermen will not reveal the exact size of their earnings at home to prevent their wives from pressuring them into incurring unwanted expenses.

Upon entering the houses of high-producing boat fishermen, I got the impression that their living standard seemed on a par with that of the canoe owners, even though only one boat fisherman owned a televison set as opposed to five canoe owners. However, when I accompanied them to the bars and noted the money spent, it became clear that their household budgets had to be considerably lower than those of canoe owners. The *vale* is sufficient to feed and clothe the family but leaves little leeway for variation and quality. Wives of high-producing fishermen will therefore try to earn some money as charwomen or seamstresses to buy food instead of spending everything on clothes and kitchenware. Boat fishermen do not deliberately neglect their wives and children and they are, of course, not insensitive to illness, hunger, and suffering.[9] Still, they are not intimately familiar with the household and acknowledge that they prefer to be far from the many day-to-day demands and the domestic preoccupations. When they deliver the *vale* at home and leave the harbor, they also leave their public and domestic roles behind, and concentrate on the catch to come. They learn consciously to eliminate troubing thoughts about home so as to maximize their fishing effort, as is illustrated by the following poignant exchange between a boat owner and a boat fisherman.

> The crew had gathered in a bar to receive their meagre shares from a ten-day fishing trip ruined by bad weather and a small catch. Moreno had earned Cr$15 ($0.03), not even enough to buy a beer. "You know you still owe me five thousand cruzeiros, eh Moreno?", said Serafin as he finished calculating the catch division. "Well," replied Moreno softly, "you know that I always try to do my best but I was demoralized." "Demoralized?", said Serafin with surprise. "Yes sir, I could not sleep because I was worrying about my woman who is about to have her first baby." "Listen," admonished Serafin, "all fishermen go through this during their first year of marriage, but at sea you have to think only about fish, fish, and more fish. Don't worry about

> your woman being in need of money, she will struggle her way out" (*ela se vira*). Serafin followed this subtle allusion to illicit sexual behavior with the explicit statement, "If your wife turns you into a cuckold, then you make a cuckold out of her" (Se sua mulher bota chifre em Você, Você bota chifre nela; literal meaning: If your woman places horns on you, then you put horns on her).

Several times I have heard boat owners give this advice to inexperienced fishermen. Economic and domestic reponsibilities are disengaged, the conjugal relationship is instilled with mistrust, deceit and disloyalty, and indifference is portrayed as beneficial.

The household budget is seldom a subject of contention between boat owners and their wives. Boat owners amply fill their houses with luxury items, and always give enough housekeeping money to their wives even if they do not go to sea. Unlike canoe owners, these absentee boat owners do not want to be bothered with the daily finances of the house.

The financial situation of poor boat fishermen is so precarious that the household budget is a major source of discord. The budget is, at best, equivalent to the minimum wage—if the *vale* were to be paid out regularly and passed on to the wife. However, boat owners do not like to extend too much credit to already indebted fishermen when a fishing trip is delayed. Marital conflict is greatest during these prolonged waiting periods between trips. The wife questions her husband's masculinity, and he barely protests because he knows that he does not measure up to the ideal self-image of a man who provides well for his family and can confine his wife to the house. His honor and respect as a man have been tarnished (cf. Gudeman 1978:41). The woman becomes the actual head of household and creates a matricentric family (Azevedo 1965:298). She assumes the parental responsibilities of her husband. Women in such circumstances appropriately call their house a "women's house" (*casa de mulheres*).[10] As one woman told me, "The women in Camurim cover the mistakes of the men."

> Sauza's subjugation at sea and his marginalization at home led him to violence and abuse towards Dalva. Being a woman he believed Dalva to be inferior to him, but by being able to support the household she defied her dependence on Souza, reversing male-female role conceptions, and thus adding to his feeling of inferiority. Dalva tells about his bad temper, "He would barge into the house at midnight and ask for fish stew. Then after twenty minutes he would wake up and ask me where his plate of liver was. And I would say, 'but Souza, you wanted me to prepare fish.' And then he would throw his plate on the floor and hit me. At other times, he demanded to sleep with me at the moment I had to go to work. It's one thing after the other. Last time, I had his clothes all ironed, and then he threw them on the dirty sand, held them in front of my face and asked me why I hadn't washed his clothes. And I just looked at him, Senhor Antônio, and did my duty as a wife."

Canoe fishermen who switch to boat fishing begin to display much of the same behavior as high-producing fishermen. The long fishing trips made Enrique less aware of domestic needs. His economic universe was no longer encapsulated by the domestic world. He had become absorbed by the exigencies of the life aboard ship. The social, spatial, and temporal distance between production and consumption, between the house and the sea, and between labor and its emotional gratification had become too large. Enrique paid all outstanding debts promptly but he began to regard as superfluous, costs that used to be considered justified expenses. Unable to feel respect from family members and former colleagues because there were fewer opportunities to reinforce his self-esteem, he immersed himself in the public world. When he met canoe fishermen, he nonchalantly cast around the names of important townsmen he had become acquainted with and boasted about the offers of financial backing to buy a boat. Less money arrived at the household and relations with his wife became strained. Enrique expected his wife to make ends meet with the *vale*, while he spent the remainder in public places.

Consideration for domestic needs turned into obligation, and the low *média* made him feel even more removed from the household.

Boat fishermen who temporarily switch to canoe fishing quickly change their conduct to social patterns comparable to those of canoe fishermen. In the beginning, Vítor's presence at home led to tensions. He had been unaware of the daily routine in the house and he wanted to make changes that were not to Marlúsia's liking. He tried to asert himself as the head of the household and wanted to delegate his wife to a secondary place. However, his presence also entailed a greater intimacy with domestic problems and gave Marlúsia opportunities to coerce him into assuming more responsibilities. The overall effect of his switching production modes was so positive that Marlúsia urged Vítor to continue canoe fishing.

## Domestic Autonomy and Disparity

Canoe owners and canoe fishermen who direct their economic efforts to the household do more than respond to domestic needs, they sustain the house as a cultural focus, as a focus of activities and cultural meanings central in their lives. The domestic world is regarded as standing above other realms of society because the house is not an asset to be mortgaged for economic gain, the household budget is not determined by a capital owner but rather directly reflects the earnings of the day, and the moral reputation of the family is not endangered by an institutional threat of adultery. The canoe owner and canoe fisherman are in complete control of the household's economic basis because of their independence as producers. They do not have the financial security of prolonged credit but neither do they run the risk of indebtedness, and the household will not be incorporated into the production mode through the extension of economic obligations into the domestic world.

The noninterference of others in the domestic unit gives the house its autonomy. The canoe fisherman, his wife, and their children create a world on their own terms, with social practices that are established by them (cf. Berger and Kellner 1964). Even though the male head of household has more power and authority than his wife, the intimate face-to-face encounters and the continued presence of all members give women more opportunities and power to renegotiate the cultural practice of male dominance than those who are married to boat owners and boat fishermen. This control of the domestic world makes the house a private, almost sacred haven from the oppression, exploitation, and patronage that are rampant in the highly stratified society. The loss of this control is always imminent and stimulates canoe fishermen to work hard to retain a fishing mode that secures their independence and ensures their autonomy.

The autonomy of the household is strengthened by unanimous agreement among the household members about their objectives. The house is an expression of unity because the members share similar goals, namely, protection of the family's moral integrity, raising the standard of living, interacting with intimacy, displaying the advancement of these objectives to others, and as an important but still secondary consequence gaining prestige in the community.

It is not by accident that the interpretations of the social practices of the canoe and the house overlap. The interpretation of a cultural focus may bridge several social domains. If the practices of the cultural foci are congruent, then the interpretations will adjust toward a greater compatibility that, however, will never be perfect but will always be in process. The canoe fisherman's emphasis on the independence and autonomy of the household will direct him to economic actions which will support those interpretations. Corporate groups, a daily routine of fishing, avoidance of bars, and negative stereotypes of boat fishermen serve to prevent a switch to boats. The emotional satisfaction of securing and improving the economic base of the household will add to the significance of the house. The public domain will be relegated even further to the periphery of the fisherman's universe and stimulate him to improve the household's standard of living.

Boat owners and their wives do not interpret the house in this way. The household is marginal to their economic and public interests and the house appears to be little more than a place to rest. They do not seem to sense the unity perceived by the canoe fishermen. The women hold the same disdain for manual work as do the men and yield the intimate routines of childcare—which would help to bond the relationship of mother and child—to a nursemaid (Iutaka 1971:206). Husband and wife both pursue public interests, even though the wife attains these through the house.

The households of high-producing fishermen are divided by a fundamental conflict between husband and wife. A woman who compared her plight to the harmony in the households of farmers and canoe fishermen lamented that boat "fishermen and their wives live more apart, because they follow separate ways." The man's public objectives run counter to the woman's interest in the household. The women regard the house as their female territory and as a symbol of womanhood. The house is an extension of the self and family life is the realization of motherhood.

Most women interpret the house as a symbol of womanhood but other meanings may overshadow this symbolism. The wives of boat owners see it as a status symbol and the wives of canoe fishermen adhere to the male interpretation. The wives of poor boat fishermen have an opposite interpretation. The house is characterized by its absence of unity, its internal strife, neglect, and destitution. The house symbolizes the woman's struggle to sustain the family without her husband's help. It also stands for her triumph over the unequal odds allotted to women in a male-dominated society. The "women's house" symbolizes her achievements despite her subjugated role as a woman.

# 8
# Public Image and Social Prestige

Joaquim entered the bar around noontime. Geraldo the boat owner, Isidoro his captain, and the other men in this group of friends looked at him coldly. Joaquim had fished for several years in Geraldo's crew and was among the best fishermen in Camurim. Yet his excessive drinking was often a cause of considerable annoyance to other fishermen. Several months ago, Joaquim had entered into a vehement argument with Isidoro and was immediately discharged. Joaquim sat down. Nobody offered him a glass of beer. His dismissal from the crew had made him fall out of grace with Geraldo's drinking group, and this afternoon he soon realized that he was still unwelcome. Agenor initiated the public humiliation. As he was leaning against the doorpost behind Joaquim's back, he reached out and grabbed Joaquim by the ear. "What a flabby ear!" he exclaimed. Isidoro repeated Agenor's gesture and said, "What a flabby ear! . . . Who has a flabby ear? . . . A rat! A rat eats cheese. Cheese comes from a cow. The cow belongs to the bull. The bull has horns . . . he has horns! [he is a cuckold]." The men roared with laughter. Joaquim stared stoically in front of him. "What's the matter, Joaquim?" said one. "He's thinking up an answer!" said another. "If you are a rat," resumed Isidoro, "then you are a thief." Joaquim stood up and left the bar without uttering a word during the entire exchange.

The mortification of Joaquim in this incident demonstrates the volatility of social relations in bars, and the intertwinement of the public and the economic domain. Fishermen may lose their employment, can be cut off of an entire cluster of contacts, have promising mobility channels shattered, and their hopes frustrated, simply by falling out of favor through a haphazard squabble. Careers are made and broken in the bars of Camurim. The bar is the cultural focus of the public domain which provides boat owners and boat fishermen with a context for voluntary interaction, the manipulation of social relationships, and the formation of networks.[1] Barroom interaction in Camurim deserves close attention not only because much time and money are spent there, but because it is in the eyes of the boat owners and boat fishermen part of their economic practice.

Chapters 6 and 7 demonstrated that the domestic domain does not take a prominent place in the lives of the boat owners and boat fishermen of Camurim but is treated as a bounded domain with limited demands. Unlike the canoe fishermen whose economic success is expressed in the domestic sphere, boat owners and boat fishermen have to transform their economic achievements into public acclaim. The boat owners and boat fishermen organize their economic practices around the demands of the bars and brothels they frequent during their long weekends ashore. Both groups will evaluate the success of their production efforts by the ability to demonstrate their generosity and masculinity in public, and by the awards of social prestige and advancement. These social objectives are part of the context of their economic practices, and relate the economy inextricably to the public domain.

In certain ways, the bar resembles the house with its various social spaces and forms of social interaction. Camurimenses distinguish between four types of bars that are characterized by unique social hierarchies based on different ranking criteria. People have to prove their worth in each setting because the indicators of social rank, such as masculinity, prestige, honor, manners, and income are interpreted differently in each social context.

In this chapter, I will begin with an explanation of the contradiction between the social differentiation pursued in bars and the paradoxical claim of equality made by the customers. Social interaction in the public domain is supposed to be voluntary, unconditional, and free of the status differences that reign in other social domains. Behind this thin guise of equality lie obligations, responsibilities, friendships, alliances, networks, and social standards that place customers into social hierarchies and ranking systems. After classifying the bars of Camurim, I will proceed with a detailed discussion of the patterns of interaction in the four types of bars. The final section recapitulates the relationships between the economic and the public domain through various interpretations of the cultural centrality of the bar. What goes on in bars is not isolated from the economic domain because economic concerns are often concealed below the surface of trivial chatter and carefree toasting.

## *The Paradox of Equality*

Rosalvo's bar is a popular halting place for fishermen. Located in sight of the harbor, the bar's large open doors invite passersby to step in for a moment and chat with its owner. Rosalvo often sits on the counter leaning with his languid body against the wall as he listens attentively to his customers. Aside from rows and rows of bottles of all kinds of liquor and a picture of Iemanjá, the place is empty. "Rosalvo," I asked him once directly, "why do fishermen frequent bars?" He merely pointed his finger to a handwritten sign on the wall: Espero que nunca nos meus olhos, e nem sintas nos meus gestos, vontade de ser superior á ti (I hope that you will never see in my eyes, nor feel in my gestures, the desire to be superior to you). "In a bar, everybody is equal," he explained, "I put that on the wall so that nobody will forget it." A couple of blocks from Rosalvo's place, in the bar where Joaquim was humiliated, the owner had posted a similar sign: "O homem por mais ter [sic] inteligente que seja, não abusa da consciência dos outros, assim ele é digno da inteligência que possui" (No matter how intelligent a man is, he does not take advantage of the conscience of others, in this way he is worthy of the intelligence he possesses). Clearly, bars are supposed to be neutral grounds where status differences are suspended and customers treat each other as equals.

Yet, although equality may be wished for, people will inevitably distinguish themselves by what they drink, how much they spend, how they talk, with whom they associate, and how they are dressed. The pretense of equality does not restrict but, in a paradoxical way, facilitates the manipulation, acquisition, and display of rank and prestige. In an effort to legitimize this privileged situation of equality in a society which is stratified, boat fishermen exert themselves to rival the conspicuous consumption of the rich and powerful. Their inevitable failure to do so can be interpreted in such a way that they themselves are to blame rather than the structural imbalance concealed in the guise of equality. Boat owners and high-producing fishermen most openly proclaim the equality of barroom interaction. Privately, however, these same fishermen complain about the subtle gestures of superiority by the boat owners and the pressure to conform to a hypocritical pretense of egalitarian conviviality.

The common practice of drinking alcohol together and ordering drinks in rounds emphasize this paradox of equality. A man is expected to finish his glass as fast as the other men in the group so that no one can display his greater ability to hold liquor while, at the same time, slower drinkers do not interrupt the steady alcohol consumption of their companions.[2] This practice prevents those with little money from drinking with moderation, and forces them to spend more than anticipated. Customers who stand a round of drinks expect others to reciprocate immediately. When a person cannot pay for his companions, he should ask the bar owner for credit, borrow money from a friend, or promise to reciprocate at a future occasion. He can also leave the bar or announce that he only drinks on his own account. A retreat leads to a loss of face, while other options cast doubts on his solvency and suggest that he does not really fit in the group. Still, boat owners successfully exploit this situation of reciprocal drinks to raise their social status. After an evening of drinking they may offer in a grand gesture to pay for the last few rounds and the expenses of snacks and meals, knowing that the fishermen cannot match their generosity. The acceptance of these drinks legitimizes the power and superiority of the boat owners and the subordination of the recipients.

The paradox of equality not only exists in fleeting relationships at spontaneous gatherings but also in long-term relationships among semi-permanent groups. Men have drinking friends with whom they are on amicable terms in bars but who are not considered "real friends" (*amigos de verdade*). Camurimenses make a clear distinction between *amizade* (friendship) and *amizagem* (drinking friendship). Amizade is based on trust (*confiança*) between persons who have mutual consideration in a friendship which has been proven at numerous occasions. Amizagem, on the other hand, is based on suspicion (*desconfiança*) and is characterized by a high yet precarious *média*. Drinking friends meet in bars and help each other out when one of them is short of money. They expect to be reimbursed quickly to maintain their good understanding because their friendship is confined to the public domain and does not hold ground elsewhere. Amizagens are quickly struck, but amizades are slowly built up after years of interaction during which obligation is created and consideration nurtured. Boat owners and boat fishermen consciously try to turn amizagens with elite members into amizades so that they can draw benefits in other social domains. The assumed equality of amizagens is used to overcome the structural inequality of persons with different socioeconomic statuses.

## *A Classification of Bars*

Camurim has about thirty-five bars and at least ten small stores that also serve alcohol. These ten establishments are only frequented in the late afternoon by canoe fishermen and other townsmen who buy a cup of baking oil or some spices for dinner and take a shot of rum as appetizer.

The bars of Camurim can be classified into four types in an order of decreasing public esteem and reputation: weekend bars, daytime bars, brothels, and skid row bars.[3] Each type is situated in a particular neighborhood that corresponds to the sociospatial organization of Camurim described in chapter 6. The locations of bars and domiciles are related indicators of prestige in the community. The more reputable the bar, the better the neighborhood, and vice versa. A drinking establishment which attracts a lower class clientele will not be tolerated in an affluent neighborhood out of fear for the safety and moral standing of the residents. A too ostentatious crowd might get easily out of hand, as it does in brothels, and cause brawls and fights.

Weekend bars (*bar*) have their busiest days on Saturdays, Sundays, and Friday evenings. These bars are located along the former and current main street of Camurim, and are the only places that carry proper names displayed on large signs above the entrance. Weekend bars are decorated with mosaics and paintings, have small round tables, play popular American music on excellent sound installations, and serve mainly beer, whiskey, gin, vodka, and other expensive drinks. *Cachaça*, a Brazilian sugar cane rum, is not served but customers can order cocktails (*caipirinhas*) made with brand name Caribbean rum. Boat owners, a few high-producing boat fishermen, and members of the elite form the clientele of the weekend bars. The customers are well-dressed and are often accompanied by their wives.

The harbor is laced with daytime bars (*venda, balcão*) which are open from sunrise till sunset from Monday through Friday, on Saturdays till mid-afternoon, and on Sunday morning during market hours. These bars sell beer, cachaça, and groceries. The groceries are often sold on credit to crews and the wives of boat fishermen. A large wall-to-wall counter which prevents customers from reaching for the wares dominates the place. The walls are bare except for a calendar and an image of Christ. The clientele consists mainly of boat owners, boat fishermen in their work-clothes, and an occasional customer who makes a quick purchase.

Camurim has three barrooms that are known as brothels (*bruega*). These are not commercially operated houses of prostitution but resemble men's clubs in which boat owners and high-producing boat fishermen assert their masculinity and meet prostitutes who wander in to pick up customers. Cachaça and herbal liquors (*jurubeba, catuaba*) are served occasionally but beer is the most common drink. Late at night, snacks and sometimes even elaborate meals are prepared to accompany the drinks. Each brothel has a small group of loyal customers to which it is difficult to gain access. When most members of the group are ashore, the brothel opens during the late afternoon and closes only well after midnight. The place has a wall-to-wall counter, a few tables and chairs, and a cheap record player that plays popular Brazilian songs. The walls are adorned with images of the Virgin Mary, Iemanjá, and Saint George (the Christian counter image of Oxossi, the Goddess of the Hunt). In addition there are a half-dozen bars which are not visited by prostitutes but which provide an atmosphere similar to the brothels. These evening bars evoke the same social comportment among the fishermen, and are included in my discussion of adultery, masculinity, and prestige. If these bars are located far enough away from the main street, then they may eventually become meeting places for men and prostitutes.

Brothels and skid row bars are located in alley ways off the main entrance to Camurim. This neighborhood is inhabited by poor salaried workers. Skid row bars (*boteco*, *botequim*) are only frequented by low-producing fishermen and other poor people. These small bars have a desolate atmosphere. A naked electric light bulb illuminates the place, which has a dirty counter. An outdated calendar is its only decoration. Cheap, locally distilled cachaça is the only drink available and the few glasses go from mouth to mouth. Skid row bars are open seven days a week, from noon till late at night. The customers are poorly dressed and often drink in silence. The clientele varies periodically when men run out of credit, and move on to another bar until they can pay off their drinking debts.

## The Transition from Economic to Public Status

After the catch of Geraldo's boat had been loaded into the truck, the crew strolled to Besouro's bar. Walking side by side, the men discussed their earnings. "I guess that with the two large groupers I'm already going to pay for my share of the costs," said Lorival. "It all depends on whether or not Geraldo will include the anchor we've lost and the cooking pot we've busted," replied Agenor. Besouro was expecting the men and had loaded his refrigerator with beer. "Beer and a bottle of cachaça," ordered the fish dealer. He took a large sheet of gray wrapping paper and began to make his calculations.

The daytime bar is a transitional space between the sea and the land, where boat owners and boat fishermen undergo a social transformation from producers to members of the community. Although these two social groups experience different status changes, they express the transition collectively in the bar. The formal and informal hierarchical labor relations are suspended ashore and alcohol is consumed in a collective spirit of equality. The men reemerge as persons with unequal public statuses after the catch division has been completed.

During the hour and a half period in Besouro's bar, the men engaged in a lively conversation, and beer and rum flowed freely. This time spent together is very important in dissolving the social organization of the crew at sea and reuniting a group of people whose intimate association during a week of discomfort often caused irritation and tension (cf. Nolan 1976). The exchange of drinks restores brittle labor relations and helps to smooth out conflicts. Unresolved differences might disturb the next fishing trip and embitter the stay ashore.

Beneditto took a bottle of beer in both hands and used his teeth as an opener. The crew did not behave like a group of drinking friends who drink in rounds with each participant holding onto his own bottle but Beneditto poured beer in all the glasses standing on the wooden counter. Aside from the dealer and the boat owner, the men were dressed in torn workclothes and no distinction of rank or status was made. Isidoro kicked Batista, and both laughed in good humor. "So, Batista, are you still thirsty? A couple of days ago you said you were only going to drink a glass of nice clean water and now look at you, already with foam on your mouth." "Listen," replied Batista, "I'm sick of drinking that lousy river water, why can't we just have someone bring tap water aboard?" The men engaged in frequent physical contact and embraced each other after a minor dispute had been recalled and forgiven. "I became really angry with you when you started casting three fishing lines at once," Beneditto complained to Agenor. "You must give us also a chance and not only think of yourself." "What happened?" I interrupted. "Well," elaborated Beneditto, "he cast too many hooks and my lines became entangled. I felt like killing him. We lost at least twenty minutes trying to unravel them." "Listen, Antônio," said Agenor defensively, "when there's fish you've gotta take advantage of the situation . . . but next time I'll take better care. Let's forget it." "Eh, let's do so," agreed Beneditto, and he filled a glass with rum. He drank half and handed the rest to Agenor, who finished it. They patted each other on the backs and checked on the calculations made at the counter.

Fishermen need a smooth working relationship to help each other secure large fish and prevent serious accidents with sharks, stingrays, and barracudas. They also want to be given fish when they are ill or injured, and they like to receive tips on vacancies among other crews. Drinking keeps these social networks intact.

Geraldo, the boat owner, also likes to be on good terms with the fishermen. He does not want to lose a good crew, and the fishermen do not want to be discharged for a trivial disagreement. Men have quick tempers and remnants of irritation may boil for days until they explode into insolvable conflicts. Captains have abandoned crews at the moment of embarkation, fishermen have stabbed and, allegedly, on one occasion even killed, a colleague at sea. Captain-owners have discharged fishermen without prior notice and left their boats in the harbor for weeks till the disputes were settled.

Canoe fishermen who switch to boats often do not join the group after their first trip. They go home and visit the boat owner hours later to receive their pay. The owner inquires about their absence, jokes that he has no money left, while accusing them of not being loyal to the crew. The canoe fishermen understand the message and will certainly accompany the crew after the next trip. Alcohol consumption has become a necessity to maintain their position on the boat.[4]

> After all accounts had been settled, bets were paid up. Lorival had promised to catch more fish than Agenor, but had lost. However, not Lorival but the victorious Agenor immediately purchased two beers and a plate of snacks. Batista was also in a good mood. "Finally, I've paid off my debt!" he exclaimed. "That's right," said Geraldo, "from now on you're going forward." "Sure," replied Batista enthusiastically, "next time I'll do even better. I know I can do it. Besouro! Get us some more beer."

The net effect of these exchanges is a leveling of incomes and a reduction in the revenues for the households of low-producing fishermen who already live on the edge of subsistence. Still, high-producing fishermen such as Agenor buy drinks for others more often, and this leveling allows them to emphasize their higher status at sea and ashore. The feigned attitude of equality consolidates rather than dissipates the ranking system among the boat fishermen.

The principal activity that brings the men together is the redefinition of their public statuses. The division of the proceeds is the cumulation of the production process and this income distribution indicates whether the economic statuses of persons have improved, deteriorated, or remained stable. A bad catch drives a poor fisherman further into debt, an injury temporarily postpones an ambitious fisherman's chance to buy a small boat, and a lucky catch makes a young fisherman prove himself a worthy addition to the crew. One boat owner has to replace a blown engine, obliging him to borrow money from the bank; another decides to sell his catch to a new dealer who offers a higher price; and a third fails to repay a dealer's advance. These changes are not dramatic and do not result in persons moving from one social status to another, but they are significant enough to guide the everyday interaction during the days of leisure ashore.

> Rúbi did not make any money. He received part of his advance and made some small food purchases at the daytime bar before going home. Lorival and Agenor joked about his poor diet, further enhancing Rúbi's feeling of shame. The poor living circumstances encountered at home would disillusion him even more. Batista, who fared better, will be reminded frequently of his good fortune during the days to come. "Batista," said Geraldo as he leaned on the fisherman's shoulder, "why don't you come to Palmeira's bar tonight and buy us some drinks?" This invitation to the boat owner's favorite brothel kindled Batista's hope of improvement and seemed to confirm the principle of equality by which interaction in public places is supposed to be governed. At least for a few days, Batista had regained the respect of his peers.

Boat owners and high-producing fishermen will generally avoid any association with their poor colleagues ashore. They accentuate their higher status by attending brothels which have a strictly controlled clientele. Again, these social groups change slowly in composition through the catch divisions. One high-producing fisherman may be informally promoted to the inner core of boat owners, while another whose economic performance has been declining steadily might be treated with more reserve.

After the revenues have been paid, the men shed their temporary equality and common identity as "seamen" (*homens do mar*), and emphasize their ranking and social differentiation. The transition between the sea and the land has been completed and the men are now members of the community. They go home, wash themselves, change into fresh clothes, and eventually merge with social groups that represent their public status and interest.

## *Adultery, Masculinity, and Prestige*

Marriage is a demonstration of masculinity. A fisherman is not considered a man when he is single. With the exception of some widowers and a few divorced men, all fishermen and capital owners have stable female partners. A married man shows that he has strong sexual needs, has the capacity to produce children and maintain a household, and can dominate his wife through a dependency relation in which he takes all major decisions.[5] Preferably, his wife is not employed so that he can be in control of her public activities.

Once a man has established a household he has to give his peers other proofs of masculinity to maintain his respect as a man.[6] These recurrent demonstrations vary with the social group in which they are displayed. Canoe fishermen look for evidence of masculinity in the economic and domestic domain. A person is macho when he has great physical strength, takes good care of his family, and dominates his wife. "We can row for hours and hours on end without stopping," claimed Casimiro. The other men nodded in agreement. "I would like to race one of those cuckolds to Pedra Grande, and let's see who wins. They're like women, they don't have any strength." Boat fishermen are discernibly less muscular but they brush off any doubts of their masculinity. When I told them of the canoe fishermen's opinion of them, Zé Silva was indignant. "That's ignorance! Those Indians from the north use their brute force, while we use the advances of technology. Which is stronger? The tree or the axe?" They act out their masculinity with powerful motorboats, as is clear from the ways boat fishermen like to race their engines and rapidly accelerate the vessels in the harbor (cf. Gilmore 1987:142).

Canoe fishermen relate strength to economic performance. A person who works hard will have a reasonable income—a correct assumption in coastal fishing where daily catches are small—and can provide well for his wife and children. A high standard of living is a more important sign of masculinity than strength and income because these last two can be wasted without adequately supporting the family. A high standard also implies that a man's wife does not need a job. Fábio, the canoe owner, consciously differentiated his masculinity from that of the boat owners. "They say that they are men because they can drink a lot. But that does not make a man. A man like that is a man of the bottle. There are boat owners here who do not even own a house. Like dogs, they live in the street. What makes a man a man is his character in treating his family well."

Physical strength and domestic comfort are less important among boat owners and boat fishermen. They go to sea in motorboats, weigh the anchor, and set the sails jointly. Of course, when fishermen pull an 80 to 100 kilogram shark on deck, they receive credit with remarks like "macho, hem?!" but such effort is not given too much importance. Likewise, the standard of living is not entirely ignored, because it demonstrates the ability to take care of routine obligations. Yet, boat fishermen generally do not visit each other at home and cannot use the domestic domain as a place for the display of masculinity. Boat owners and boat fishermen show their masculinity, therefore, through adultery, drinking, and conspicuous consumption in brothels.

Given their emphasis on masculinity, it is surprising to hear boat owners and boat fishermen say that "all fishermen are cuckolds." With this common saying, they refer also to boat owners who command their crews at sea. This open admission, which is seldom based on real evidence, seems strange in a culture which places such great emphasis on male honor and female fidelity (Azevedo 1965; Willems 1953). However, the potential unfaithfulness of women during fishing trips makes men take this assumption for granted. "You know, Antônio," explained Raimundo, "it's not because we cannot dominate our women but it is the fault of boat fishing." Adultery in brothels is a way to prove their masculinity in view of the impossibility of being sure of their wives' fidelity. Sexual prowess shows that they are machos (Gilmore 1987:126–136). They brag about their potency through anecdotes of their voracious sexual appetites and make sure that friends acknowledge their inflated claims. All fishermen may be cuckolds, but it is because of circumstances imposed by the production mode, not because of their lack of virility. Frequent sexual contact with women in public places negates, neutralizes, and obliterates any doubts about their masculinity.

Indebted boat fishermen have much more reason to be concerned about both their wives' infidelity and their own image as men. These fishermen, however, are not welcome in the three brothels of Camurim but visit women who receive men and adolescents for a small sum. Still, the low public image of poor fishermen is not enhanced by these contacts and they are not considered machos by other men.

Canoe fishermen never proclaim that they are cuckolds. On the contrary, to call somebody a cuckold is a very serious insult. Their continued presence at home is not supposed to give women the opportunity to engage in extramarital affairs. Canoe fishermen only go to brothels when there is a major conjugal conflict or when they switch to boats.[7] But even then, they seem to be troubled by feelings of guilt. Amerino complained about difficulties in completing the sexual act when he visited a prostitute. "While I was lying on the dirty sheets with that woman, I kept thinking about my wife, the children, and our home. I became temporarily impotent. When you go to a brothel, you have to keep your family off your mind."

Boat owners and boat fishermen explained that during the first years of marriage, they felt uncomfortable when their wives complained about the sexual escapades. After some time, they learned to separate the house and the brothel as different domains with different standards of sexuality and affection, while their wives began to tolerate adultery as long as they were well taken care of (cf. Willems 1953). This acceptance served as yet another confirmation of masculinity because it showed that the men can do whatever they want despite their wives' disapproval.

All bars are owned by men, but women run the three brothels of Camurim. The women do not operate as madams who receive a cut of the money earned through solicitations, but they facilitate sexual liaisons and are trusted by the prostitutes.[8] Prostitutes have often been victims of an early divorce, were unable to remarry, and finally resorted to prostitution because of social exclusion and economic deprivation (Bacelar 1982; Figueiredo 1983; Kottak 1977).[9] They are usually between twenty and thirty years of age and live in very poor circumstances in the vicinity of the brothels. They take their customers home or use a small room in the back of the brothel. The price of their services fluctuates around Cr$1,000 ($2) for an encounter of several hours and they expect to be bought a few beers and some food. Boat owners sometimes have long-term relationships with prostitutes. These arrangements may last from several weeks to a year but are almost always terminated because of the relatively high cost.

The semi-permanent arrangement with a prostitute imitates much admired elite members who have established separate households with mistresses in Camurim or in neighboring towns. A mistress enhances a man's image of being macho because he has the exclusive right of intercourse and the children are proof of his virility, a claim which cannot be made with prostitutes (Stevens 1973:59). Furthermore, the man does not have to compete with others for sexual services and can avoid the compromising low-class environment of brothels.

Boat owners associate in social groups that hang out together. Low-producing fishermen split up into several drinking groups, while most high-producing fishermen associate with groups of boat owners. The latter groups are called turmas. The ten to fifteen members of the two principal turmas get together for soccer practice (*bába;* an abbreviation of *bate bola* or to play ball), offer help with engine repairs and fishing supplies, and even give each other hints at sea about the location of fish. Still, these turmas are not corporate groups similar to those of the canoe fishermen, because they do not control territories or fishing grounds, and change regularly in composition. The core of a turma consists of a half-dozen boat owners who informally decide about its membership. A highproducing fisherman who is invited to a brothel may become part of the turma if he is liked by the core group. The turma does not want to be bothered by "intruders" (penetras) who ask for money and free drinks. The close-knit group strictly controls the clientele at a brothel to protect their privacy.

> We all looked up when Bastos staggered into Palmeira's bar one night at about eleven o'clock. He was obviously drunk. "Tomorrow I'm gonna go to Rio de Janeiro and fish on a large vessel. I'm the best fisherman in Camurim . . . I'll show you!" "Sure you will," remarked Valdemir sarcastically. "Like that catch of seven kilograms you made last month." The men laughed. Bastos became angry, calling them leeches and cuckolds. "You'd better shut up," said

> Larsi, "because you still owe me forty thousand cruzeiros." "Let me tell you something," started Bastos. "The harbor master in Salvador told me that a fisherman has a *right* to his credit, and the owner has an *obligation* to give it" [he emphasized both words]. Larsi denied this vehemently. Suddenly, Bastos launched for the table. He tried to grab an empty beer bottle supposedly with the intention of striking Larsi. The men responded quickly. Palmeira pulled me behind the counter because a fight seemed immanent. The men took him by his arms and, as they threw him out the door, Larsi hit him so hard in his neck that Bastos fell into the street. Larsi shouted after him, "Bastos, no more *brincadeira* [teasing]. There's no more friendship. No more." The next day Larsi reflected on the incident. "He came barging into our place, and then when he started talking about the *vales*, I became really angry. The s.o.b. owes me forty thousand cruzeiros. He took my blood. It's not just the money that counts, it's the blood spilled at sea to earn it that he has to pay."

Brothels are well-liked because of their exclusive and intimate environment. The brothel provides the best atmosphere for men to express their friendships and feelings. "In Palmeira's bar," explained Tavares, "We can be ourselves, without being bothered by our wives." The music enhances this feeling of togetherness. The melodramatic themes depict men as victims of their circumstances, and can put the group in a very sad and silent mood. Songs of a father who explains to his son why the boy's mother abandoned him for another man, of a mason who built a school which his poor daughter could not attend, and of the loneliness of the cattle drive evoke feelings that touch upon their daily lives. "We call the brothel '*casa de família*' (family house) or '*casa de respeito*' (house of respect) to delude others," said Valdemir. "But it also means that we feel at ease there."

Palmeira is a mother figure who is highly respected by the men. She is a stocky, middle-aged woman, with a shawl covering her head and cheap jewelry on her hands and wrists. She talks with a soft voice and listens with empathy to their stories. At times, the men carry on a slow, disconnected conversation with long pauses in which they seem lost in thought. They sit on the few chairs, staring into blank space. It seems as if, at these moments, the weight of their entire existence presses on them and they can only talk about the most mundane things in life. "It's windy," said Agenor as he pulled his coat tighter around his shoulders. The record player had been turned off and the refrigerator hummed in the background. "Winter is coming," said Isidoro. Pause. "That's right." Pause. "I'm gonna go." Isidoro stood up, and left without farewells.

Often, however, the conversation consists of humorous anecdotes and gossip. The general mood is good and the men have a feeling of togetherness, heightened by the steady consumption of beer. Rounds of drinks keep the glasses filled and an elaborately prepared dish of seafood adds an extra touch to the warm atmosphere. The sense of equality created is genuine but it is punctuated with opportunities to assert masculinity, display social differentiation, and obtain prestige. McClelland and his associates (1972:335) have established a positive correlation between alcohol consumption and a heightened sexual aggressiveness in the United States as a consequence of a culture with a strong sense of male power and prowess. These findings are well suited for this analysis of drinking in the brothels of Camurim.[10]

Alcohol consumption can get easily out of hand if a person does not know when to stop. However, the social convention of buying rounds makes it difficult to refuse alcohol because such rejection places the person outside the group and might be interpreted as hostility (Nolan 1976:81). The proper consumption of alcohol is a proof of masculinity with a double catch. The person who becomes drunk and cannot control himself is considered weak, while the person who refuses to drink seems to be stingy and furthermore admits that he cannot take more beer. A macho is a man who can support large amounts of alcohol without showing any major signs of intoxication. He is in control and generous. Boat owners take advantage of their authority to set the pace of drinking according to their own ability and use their economic assets to display their generosity. When boat owners arrive ashore, they often send a large fish to the brothel and ask the owner to prepare it. This conspicuous consumption proves that the boat owner is a macho whose generosity is hard to surpass. Generosity and the feeling of power at dominating the social environment affirm the boat owner's manliness. Masculinity, power, prestige, and sexuality all come together within one social practice. Drinking under these circumstances gives a boat owner a feeling of personal dominance, and release "thoughts of having impact on others, of aggression, of sexual conquest, of being big, strong, and influential" (McClelland 1971:815).

High-producing fishermen sometimes try to keep up with the boat owners to show that they are just as strong. Nilton used to reserve a refrigerator stocked with beer ready for the day he returned from sea. He placed so much confidence in his ability to catch fish that he could spend in advance what he had yet to earn. Although admired by his colleagues, he wasted so much money that even his employer felt sorry for him. When the public notary decided to sell his boat, he wanted to arrange a bank loan for Nilton. However, Nilton thought that the boat needed too many repairs and instead asked for a canoe and six nets. Ever since he switched to canoe fishing, he has improved his standard of living and only rarely goes on a binge.

Boat fishermen like to be part of a turma because it gives prestige and makes them more closely related to the boat owners. Such friendship may lead to an introduction to members of the elite who can advance the fisherman's career. However, the elite never visits brothels, while boat fishermen cannot attend weekend bars uninvited. Only a boat owner can serve as their broker.

## *Patronage and Social Mobility*

The elite of Camurim personifies the social aspirations of the boat owners. Whether they are cattle ranchers, cocoa planters, or landholders, the members of the proprietor class represent what boat owners want to be: capital owners who live on the returns of their investments and wield great political power. The boat fishing industry has developed only recently and many boat owners are still consolidating their enterprises. They have not yet paid off all loans and many have to command their crews at sea to raise their earnings. However, once the boat is completely theirs, they dissociate themselves from the production of fish and stay ashore to become more firmly entrenched in the networks of the elite. Only if they can become part of the networks of kinship and social alliance will they be able to gain influence in Camurim and set the local power structure to their hand (cf. Miller 1979).

Boat owners make a conscious effort—through conspicuous consumption—to show that their capital revenues are comparable to those of the elite. Yet, although the landed elite regards capital ownership as an important status marker because of the obvious economic potential, land and boats are not considered comparable—neither as a source of income nor as a status marker. Historically, land bestows its owner with the cultural heritage of a society rooted in a colonial plantation economy and ruled by a patriarchal rural aristocracy (Iutaka 1971; Ribeiro 1971;204–221; Schwartz 1985:245–254). The landowner has omnipotent control over sharecroppers and laborers unmatched by boat owners. Furthermore, land is a primary factor of production that requires far less immediate attention than a boat. As a total institution, the boat isolates the captain-owner from public institutions and obstructs his participation in the political affairs of the county. Therefore, captain-owners try to become absentee boat owners with a minimal involvement in management and production. Valdemir described to me how he first noticed that he was regarded as a member of the local elite.

> "Last year I needed a new engine for my boat. I went to the bank and applied for a loan. The next day I received a letter from the bank to appear there and talk to the manager. I don't like the manager, he's a loathsome person (*um cara nojento*), but let's leave that aside. I went over and said 'What's the matter?' He said, 'I read your proposal but . . . how many cosigners do you have?' I said, 'How many do you need? Ten, twenty, just tell me.' 'Well,' he said, 'here is the paper, and just give me two names and then you obtain the signatures.' I went out and immediately got

> the signatures of two cocoa planters. I returned to the bank and received the loan. Four months later, I decided to replace my nets. I made a proposal for twenty nets, even though the sum exceeded their limit. I thought 'Let's see what happens.' The next day the manager called me in to see him. 'Anything wrong?' I asked. 'No, nothing, just that your loan has been approved.' 'Don't I need a cosigner?' I asked. 'No, for you that is not necessary.' And in this way I became a patron."

About half of the twenty-five boat owners employ captains to run their boats so that they can stay ashore when they wish, while four to six boat owners regard themselves as absentee owners who used to be fishermen but are now making the transition to elite status. The remaining six owners have never been fishermen and are permanently ashore, even though they have not yet been able to gain access to the restricted social circle of the landed elite.

Wealthy townsmen lead a very sheltered life. They never entertain ordinary people at home, and they seldom enter stores, daytime bars, or the harbor. They consciously try to minimize their contacts with people who do not belong to the upper echelon of the community, and even succeed in avoiding people in public places by receiving preferential treatment from the officials in charge. A landowner does not wait in line but raises his hand to a clerk and is immediately led into the manager's main office.

The weekend bar is the principal place where the elite socializes with other Camurimenses. Social interaction in weekend bars is organized by clear social practices and does not have the intimacy of brothels or the hectic atmosphere of daytime bars. The fishermen are strongly aware of the etiquette prevalent among the elite, and drinking helps to relax the somewhat formal mode of interaction. "A drink gives a person the courage to talk," explained Isidoro. "Drinking, he is more of a man and it allows him to talk to these powerful men."

> Although Geraldo, Raimundo, and Isidoro invited their wives to come along on the early evening outing to the bar Costa Dourada, the couples separated upon their arrival. The

women sat at one table and the men at another. The three fishermen wore their best clothes and assumed a controlled posture that seemed almost awkward. Their wives ordered some liqueur and the men drank whiskey. The conversation was carried on in low voices and the drinks were consumed with measured sips. "Doutor Barbosa," said Geraldo as he addressed the cocoa planter, "what opinion do you have of the current oil prices?" "It looks bad, Geraldo, transportation costs will go up, prices will rise, and the inflation will grow even more," replied Dr. Barbosa. "These are very serious matters indeed," remarked Geraldo. "What will be the effect on the price of the dollar?"

Geraldo seemed eager to show that he was aware of the current economic and political events, and that he had successfully emulated upper class manners and ways of speech. He wanted to change his status from that of client to colleague, from boat owner to member of the elite, from contractual partner to friend. A month ago, he had become an independent capital owner by paying off his final installment and thus ended the formal contract with his cosigner and patron Doutor Barbosa. There will always be an intangible debt to Dr. Barbosa but the social relationship has lost its unequal economic foundation. Geraldo would like to be invited to Dr. Barbosa's home to play cards, celebrate festivities, and plan political campaigns.

A change in social status and the growth of political power coincide with a changing conception of masculinity. Although many landowners consider a macho as a man who supports two households—those of his wife and his mistress—their sexual prowess will never be challenged in public. Power, rather than virility, determines their male status. A macho acts as a patron, has economic and political power, and breaks or circumvents the law with impunity. He has ample economic resources and political leverage to dispense goods and services to his clients, and a network which facilitates access to other persons. This conception of machismo is not related in such a direct way to drinking and virility but to a strong sense of honor (cf. Pitt-Rivers 1968). His extensive contacts with public officials allow him to get around bureaucratic regulations and he will break the law if his honor has been offended. A real macho will first redress the violation so as to restore his reputation, and only then try to deal with the legal consequences. Members of the elite and several absentee boat owners have shot and beaten others without ever standing trial by exerting their influence on the local police.[11]

Landowners associate honor generally with the respect due to them as the bearers of power and high status. Respect flows from the deference with which others treat them, not from the way they behave towards others. Politicians and landowners are known to cheat peasants out of their small plots of land, to accept bribes, to withhold salaries of plantation and government workers, and so on. These dishonest practices are regarded as acts of power instead of acts of dishonor. Their respect increases when they succeed in getting away with such expressions of force. Boat owners try to make similar claims to honor and often reprimand fishermen for "not respecting" them. Yet, their ambiguous social status in Camurim allows them only to draw their respect from the most dejected townspeople.

Boat fishermen and canoe fishermen have a different conception of honor as respect. They relate respect to virtue and moral character. The difference between male and female respect as interpreted by Fábio, the canoe owner, illustrates the contrast between elite and common conceptions of honor.

> "When a man is on the ground, we say: 'Get up, macho!' But when a woman lies on the ground, we say: 'Get up, whore!' Only the names 'thief' and 'homosexual' stick to a man. Only these two names can make a man lose his respect as a man. He may get drunk, fight, or hit his wife, but he can always regain his honor. But if he is a thief or homosexual then he loses his honor and respect as a man forever. . . . Not so for a woman. If she is drunk, she has lost her virtue. A drunken woman is a whore. This is not true for a man. He may drink for entertainment or for influence. The mistakes of a man are only one [i.e., adultery]. The one he is entitled to. The woman does not have a right

to her mistakes. The rights of a woman are to take care of the house and to respond to her husband's demands."

The social mobility of boat owners is a complex process that involves changes in appearance, manners, networks, behavior in public places, and their conception of masculinity. Boat owners will eventually have to withdraw from the disreputable environment of brothels and prostitutes. These public places make them vulnerable to demeaning requests and damage their image as morally upright men. However, they can have mistresses as long as these are loyal to them and keep a low profile in public.

During the early days of the fishing industry, members of the elite could be easily tempted into patron-client relationships with gifts of fish, political loyalty, and good recommendations. When several ranchers suffered the consequences of their clients' default, they retained the relationships with boat owners and helped them with small loans, but they were no longer persuaded by gifts and votes to enter new contracts. Only an intimate personal acquaintance with a client could convince them to become cosigners. Hence, the boat owner has become a crucial broker of economic mobility.

Geraldo enjoys posing as a member of the elite who introduces others to his rich friends. He will invite only those high-producing fishermen who belong to his turma, have demonstrated their ability at sea, and have proven to be loyal and trustworthy companions ashore.[12] In weekend bars, they will be very respectful to rich townspeople, sit almost immobile, and use extremely formal terms of address to express their deference.

Geraldo extolled Isidoro's capacity as a fisherman, "Dr. Barbosa, Isidoro is good with a fishing line. The piece of *canapú* you received last week was caught by him." "Very good, Isidoro," said Dr. Barbosa, "that fish was delicious." "He is very determined to own a boat himself, Dr. Barbosa," continued Geraldo, "he's working hard to earn enough money." "Very well, very well," said Dr. Barbosa evasively. "Listen Geraldo, can you get hold of some lobster? I have some important people visiting me Sunday."

"Certainly Doutor," replied Geraldo. "Don't worry, I'll take care of everything."

Isidoro was silent during the entire conversation but he had had his first formal introduction to Dr. Barbosa as an aspiring boat owner. It still might take several years before he can buy a boat. The initiative will most likely come from Geraldo who wants to replace a smaller with a larger vessel. Geraldo will negotiate with Dr. Barbosa about a bank loan for Isidoro, and he will implement most of the financial transactions. Isidoro will buy the vessel, and Geraldo will secure a new loan with the cash sum.[13]

## *Social Marginality and Interstitiality*

This chapter has described many instances of alcohol use but has not addressed the problem of abuse. Alcohol abuse is a socially and culturally defined form of drinking which does not necessarily entail alcoholism, the physiological and psychological dependence on ethanol. Problem drinking in Camurim is uniformly regarded as a personality disorder, as a weakness on the part of the drinker. However, the identical course of increased alcohol consumption and diminished productive activity through which alcohol abusers pass, strongly hints at social causes of abuse.

Camurimenses generally agree that foregoing meals to consume alcohol instead is the first sign of problem drinking. Boat fishermen need time to recuperate from the sleepless nights and the irregular meals at sea. Excessive drinking lessens the appetite and weakens the body. The case of João do Leo is typical. Several years ago, João do Leo often used to embark drunk and sleep off his hangover on the way to the fishing grounds. João became so unreliable a fisherman that he was discharged by the boat owner. He had difficulty finding another crew and ended up on boats which sailed sporadically because of frequent mechanical problems and undependable crews. He made fewer trips, spent more time drinking, and eventually stopped fishing entirely. Because of the respect and friendships built up over the years, his former colleagues gave him fish and boat owners arranged small jobs for him in the harbor. However, his incessant drinking made him lose the strength to carry the heavy boxes of ice on the slippery riverbanks. Fishermen began to make fun of him and boat owners refused to give him work. João do Leo's respect in the community was gone, he had lost his friends, and children insulted him in the street.

In order to maintain his drinking habit, João began to fish at the beach. After his wife and children abandoned him, he even lost his place on the regular beach seining crew. I first met João as a member of the shore crew. He would stand among the women and old men who pulled the net ashore, earning a few fish for his efforts. At around eleven o'clock he would go to a skid row bar to exchange the fish for a drink. João do Leo had become an object of scorn, a hilarious break from the daily routine.

> "*Opa* João, taking a walk, eh?" shouted Rosalvo. João stopped and looked at Rosalvo with his mouth open, and his shirt half tucked into his pants. His face was red and swollen. He looked much older than 43. "You want a drink, João?" said Edivaldo. "Here, here's a drink." João entered the bar and reached for the bottle of beer. "No, this is not for you," protested Edivaldo, "this is not strong enough. Rosalvo! Give me a glass of pure alcohol." I began to feel uneasy with the whole scene and told him to stop, but João said that he could stand his liquor. He gulped down the ethyl alcohol as a last proof that he was not an alcoholic but a man who could drink without showing signs of intoxication. "*Opa*, what a macho man!", said Edivaldo sarcastically. João looked at them proudly. "Good, João. Now, leave us men alone." João mumbled something incoherent and left the bar.

Some problem drinkers eventually die from cirrhosis, or accidentally drown in the river, but most of them recover after a physical collapse either at home or in a hospital. Many have gone through a period of drinking and some have repeated the pattern several times. The cycle can be interrupted at any point and does not have to run its full course before the process can be reversed. Social control, social exclusion, ridicule, and self-awareness may make a person decide to cut his destructive habit.[14]

Low-producing fishermen are not yet as incapacitated as João do Leo and they never need reach his stage. Still, their alcohol consumption also borders on abuse, if not of themselves, then—at least—of their dependents. Indebted fishermen drink to escape the anxiety of economic, public, and domestic marginality. Rúbi has the lowest rank in Isidoro's crew; he occupies the least favorable position on deck; and he is deeply in debt. The value of his labor has been insufficient to compensate for the running costs and depreciation of the boat. His indebtedness characterizes Rúbi as a low-producing fisherman who is dependent on advances partially generated by the labor of others. Economic misfortune also implies public marginality. Rúbi is excluded from important networks, social contexts to affirm masculinity and prestige, avenues of economic mobility, and nuclei of communication. In addition, Rúbi fails to maintain the household, his credit is beyond his control, and he shuns the family's social life. Ashamed to come home, Rúbi escapes to a skid row bar and drinks away his anguish.

The heavy drinking among low-producing fishermen to reduce their anxiety ashore is different from the tendency of most captain-owners and boat fishermen to lessen the stress of fishing (cf. Heath 1975:42; McClelland 1984:341). Drinking make the fishermen forget hardships at sea in order to facilitate enjoyment of the fruits of their labor and to integrate with the social world ashore. Low-producing fishermen, instead, want to escape their anguish of feeling socially marginal and of being confronted with their own inadequacies.

Skid row bars are at the interstices of the three principal social domains of the fishermen of Camurim. These bars are places where poor fishermen escape from everyday reality into a state of intoxicated oblivion. Skid row bars are public places, in the sense that they are freely accessible but the interaction at hand has no significant social consequences. Social relationships are not manipulated to attain well-defined ends, and contacts are feeble and fleeting. The transactional content of the interaction is without much substance, and drinking companions can be substituted for others without any repercussion on the social positions of the clientele.

In skid row bars, poor fishermen become almost nonpersons, devoid of any status of structural importance to the community, the family, or the economy. They do not enact their social identity as fishermen, political constituents, fathers, sons, brothers, and husbands. These social positions are of indirect importance in other bars because they allow customers to place each other in appropriate social categories, but they are insignificant in the skid row bars. The status of these fishermen is that of customers whose consumption of alcohol has no purpose beyond the psychological and physiological effects of its use. They shed all social responsibilities, and the futility and insubstantiality of the interaction reduce their status to a mere physical presence. Not even the masculinity of the customers matters anymore. Drunkenness is an expression of their disregard, contempt, and even defiance of their many obligations. In a state of intoxication, they vent their anger and dominate the street, the house, and the harbor with repetitive monologues and uncontrollable behavior. They stop pedestrians to ask for cigarettes, hang out near the entrance of public places, and talk on and on about last week's soccer match, the corruption of the government, and the dishonesty of the boat owners. But, although the drunken fisherman loses his self-control, others will still consider him responsible for his behavior. Improper behavior is tolerated and excused during drunkenness, but is neither easily forgiven nor forgotten. Once sober, the fisherman has to bear the criticism of the community in whose eyes his social status has lowered even further.

## The Bar as a Public Forum

The interpretation of the bar as a cultural focus is closely related to the discourse of boat owners and boat fishermen about the economy. Although bar owners and their customers emphasize that all are treated as equals in the bars of Camurim and that customers act like brothers, everybody knows that social differences appear as much in public as in other social domains. Different objectives in life make boat owners, high-producing and low-producing fishermen perceive other economic and public realities, and bring them into conflict about the proper interpretation of the relation between the economic and the public domain.

Boat owners emphasize that their social rank is determined by a combination of their position as capital owners and their status in the daytime and the weekend bar. They consciously manipulate the various hierarchies and use one as a stepping stone for the other. Still, social mobility is harder to achieve than economic mobility. Economic status is measured through earnings and capital ownership. It can be readily identified and signals others about the proper demeanor required at public encounters. Social rank, however, is not simply the consequence of personal achievement, as boat owners like to see it, but depends on the judgment of others. Boat owners are trying to accomplish a structural reclassification. They define elite status in terms of capital ownership, income, influence, and the privilege of retiring at an early age, while the landed elite looks at prestige, comportment, interactional skills, etiquette, and upbringing. At the conquest of each status marker, the boat owners feel that the elite subtly places a new one in their path. What they lack in the eyes of some landowners is, what Bourdieu (1984) has called, "cultural capital." The boat owners are not only wanting in their cultural knowledge and sophistication but they also fail to recognize the funds of hegemony. They have not yet mastered the strategies of manipulation and dominance that make the elite retain its power and status in Camurim.[15]

The landowners of Camurim often distinguish themselves as *gente* (people) from everybody else as *o povo* (the populace).[16] In the words of one landowner, "*O povo* stands for emotion, for hard physical work with an axe or hoe, or for work at sea. The term *gente* implies rationality, working with one's head, calculating one's steps in life, and planning a future." The distinction between *gente* and *o povo* is also made by boat owners since they feel that their capital ownership has lifted them above the populace. Boat fishermen do not find the distinction between *gente* and *o povo* to be fair, but they acknowledge that practices of social exclusion have made this stratification into a reality.

Canoe fishermen have a more egalitarian conception of personhood. "There is only one people," said Francisco. "We are all *gente* and *povo*. We all come from the same blood. It is true that one may live in a better condition than another, but we are all equal." Yet, despite this ideological belief in equality, the everyday speech of boat fishermen and canoe fishermen alike is full of references to a fundamental distinction between *gente* and *o povo*. The common expression "Você é gente!" ("You are gente") is the most often used. This expression is said in a good-humored, playful spirit to flatter a person. For example, when a fisherman treats his colleagues to a couple of beers without any apparent motive, or when a teenager succeeds in buying drinks on credit, the men will praise him with the exclamation "Você é gente!" These persons are temporarily regarded as *pessoas de respeito* (persons of respect), who for a brief time deserve to be treated with deference because of their extraordinary feat.

The bar is interpreted as a public forum which offers the opportunity to articulate status ambitions and allows men to enter a higher social class. Boat owners will not be invited to the homes of the elite and be treated as equals before they overcome the remnants of earlier patron-client associations and can somehow modify the criteria used to establish elite status. Frequent interaction in weekend bars promises to be the most suitable avenue of social mobility.

High-producing fishermen do not regard the bar as a lever with which to induce structural change but they interpret it as an arena of alliances, networks, favors, and volatile transactions. They do not want to change the social structure of Camurim but just want to smooth its connections through their participation in brothels and weekend bars. They are convinced that once they are given the chance to demonstrate their qualities and make their acquaintance with wealthy townsmen they will be successful, because the impediments on economic and social mobility are interpersonal, not structural. Rank (*prestígio*) is merely the outcome of a competitive struggle for valuable social contacts.[17] Economic achievement is a necessary but not sufficient precondition for high social rank.

Low-producing fishermen participate only in public life by way of the daytime bar. They do not share the preoccupation of ambitious fishermen with social manipulation because public contacts will not reduce their indebtedness. Their interpretation of the daytime bar as a cultural focus dissociates it from the wider social consequences of the public domain and relates it solely to the economic domain. The social hierarchy in the daytime bar is almost identical to the ranking in the boat fishing mode. The hope attributed to both the bar and the boat is similar. They have the expectation that, with some good fortune, they will be able to pay off debts and become canoe owners. This desire is expressed when they reflect on their future during more tranquil, sober, and lucid moments. However, their marginal status draws them to the gloomy skid row bars. Compulsive drinking is largely an emotional reaction to the economic and domestic dependence on boat owners, and this escape from the everyday reality locks them even more firmly on the bottom rung of the economy and the community.

# 9

# Conclusion: Toward Interpretive Economic Anthropology

As the waves roll towards the beach in endless succession, the canoe fisherman waits for "the right moment" (*o momento certo*) to enter the sea. The dugout rises and falls with the force of the surf as the fisherman uses both hands to secure it by its slightly elevated stern. Suddenly, he pushes the canoe across an oncoming wave, jumps in its belly, and works his way across the surf with the use of a long wooden pole. This seemingly simple routine takes years to master. When I asked the canoe fishermen of Camurim how they recognized "the right moment" and "the right wave" (*a onda certa*), they broke into a confusion of explanations. Some said that after every six waves of increasing height there follows a smaller one, others were convinced that the wave patterns vary with the lunar cycle, a third group claimed that periodic calms in the wind smoothed out the waves, and still another group believed there to be no pattern at all. When I asked if they counted the waves or looked at their height, they answered that they did not have to because they simply "knew." After talking to many fishermen, I had to conclude that one can only learn to get across an ever-changing surf through many years of watching waves and watching others successfully surmounting them. Apprentice fishermen need years of experience before canoe owners al-

> low them to undertake this difficult and dangerous task. Dugouts have split lengthwise and even seasoned fishermen have been injured by canoes which suddenly turned on them as they misjudged the moment of entry. The only way to learn is to witness many different situations, to grow up with the waves, not to consciously try to discover their deceptive rhythm but to become older with them.[1]

In the introduction of this book, I argued against the reduction of practice and discourse either to each other or to underlying principles and structures. A study of economic, domestic, and public practices together with the discursive conflicts about their meaning called for a close attention to interpretation in the fishing economy of Camurim. The description of the pluriform fishing economy, family life, and the public arena therefore became focused on issues of discursive confrontation rather than merely conflicts of personal or economic interest. Yet, despite the fundamental discursive incongruities, the fishermen always overcame the recurrent interpretational crises by reinforcing or transforming their power relations, and by modifying, rebelling against, or subjecting their shadow discourses to a dominant ideology. In this conclusion, I will clarify the interpretive approach in economic anthropology, elaborate on the importance of cultural foci in channeling the dialectic of discourse and practice, and argue that the conflicting interpretations among the people of Camurim are part of economic practice and reveal the economy as an interpretive construction. Each social group assumes that its interpretation is a truthful representation of the economy, and will act accordingly upon that belief. This plurality of interpretations must not be ignored by treating the economy as a bounded social domain or as an organization of functionally integrated institutions. The economy cannot be assigned to a particular place in society and culture, and economic actions cannot be described or analyzed within one coherent and logically integrated economic system that accommodates the interpretations of all actors. Truth, discourse, and contradiction are bound together in practice.

Interpretive economic anthropology starts with the assumption that people create their social and cultural existence through interpretation, and that the economy provides one particular mode of revealing that existence. People have to choose how to give substance to their lives with the practices they attend to, the tasks they carry out, the interests they pursue, the emotions they nourish, and the people they care for. How they give content to their lives is not purely a matter of free choice but is strongly influenced by the culture into which people have been socialized and the modes of expression that are available to them. The economy reveals people's social and cultural existence in one way; religion, art, politics, kinship, and ecology are other modes of expressing one's existence. My understanding of economy has been influenced by the interpretation of art and technology made by Heidegger (1975; 1977) and Marcuse (1964; 1978).[2] Art and technology are distinct but related processes of creation. Both require implements (hammers, chisels, brushes), materials (wood, steel, paint), and a degree of skill. Yet the ways in which art and technology conceive of the world and manifest it to us through their works are vastly different. The way a fishing net reveals the ocean as an abode for fish is different from its aesthetic rendition in a seascape, or from the way the possessed medium at a Candomblé ritual demonstrates the presence of Iemanjá. All three reveal the ocean to us, but one is technological, the other artistic, and the third spiritual.

Technology is an important aspect of the economy because this knowledge influences the type of capital goods, raw materials, and organizational forms used in the production process. Given the sociological approach of this study, I will consider technology and economy as one mode of disclosure because they figure together in economic practice.[3] The ecological, economic, and ideological expansion of boat fishing at the expense of canoe fishing is an example of how the existence of the fishermen of Camurim became molded by economic practice and how the fishing economy revealed previously hidden or less prominent aspects of their world to them.

When fish became scarce in the mid-1970s because of the combined fishing efforts of boats, canoes, and beach seines, fishermen failed to realize that they were overfishing their local waters.

They had been raised with a belief in the unlimited richness of the sea and could not comprehend that suddenly their marine resources were endangered. They were convinced that the fish were in hiding, had learned to detect the nylon nets, or that the Sea Goddess was offended by their laxness in making the proper offerings. Instead of giving up the myth of bounty for an awareness of scarcity, the falling catches made some canoe fishermen more sensitive to the demands of Iemanjá while prompting others to intensify their exploitation with more fishing gear. The ocean was believed to be inhabited by an intelligent and cunning species under the aegis of an omnipotent deity which had to be outsmarted. It was only years later that the fishermen began to become aware of the finiteness of the sea, both in a geographical and ecological sense. The causes of overfishing were debated vigorously, and accusations flew back and forth between canoe fishermen, boat fishermen, and beach seiners. Yet by that time, the conversion from canoes to boats had already progressed so far that a return to more conservative fishing methods was considered impossible and even undesirable. Boat fishing had developed into a fishing mode more dominant and aggressive than canoe fishing. The domestic and public lives led by canoe fishermen and boat fishermen described in this study attest to the increasing social contrasts that accompanied the two fishing modes. The interests of the boat fishermen had changed so much, after they made the switch to boat fishing, that they were not concerned about the overexploitation of the marine resources but placed capital accumulation ahead of ecological conservation, and believed that technical improvements might give access to yet unexplored fishing banks even, if that meant a further depletion of the limited fish stocks. In sum, the economic mode of disclosure brought out in the open the scarcity of natural resources, an ideology of capital accumulation, an incipient and growing social stratification, and a shift of interest from the domestic to the public domain.

What distinguishes the economy from other modes of disclosure? As a starting point, I will return to technology as a particular mode of disclosure, and focus on the closely related interpretations of Ellul, Heidegger, Marcuse, and Sahlins, even though similar thoughts have been expressed by numerous other authors. Ellul (1964:142) laments that "Technique worships nothing, respects nothing. It has a single role: to strip off externals, to bring everything to light, and by rational use to transform everything into means." Likewise, "Utilitarianism . . . is the way the Western economy, indeed the entire society, is experienced" (Sahlins 1976:167), and "Today, domination perpetuates and extends itself not only through technology but *as* technology" (Marcuse 1964:158). Heidegger argues in a similar vein that modern technology treats nature and mankind as a standing reserve, as a fund to be used up and consumed. "Everywhere everything is ordered to stand by, to be immediately on hand, indeed to stand there just so that it may be on call for a further ordering" (Heidegger 1977:298). Technology is "totalitarian" (Ellul 1964:125) and "authoritarian" (Marcuse 1964:2), giving rise to a technological, one-dimensional society in which domination takes the form of the deep force relations that have also been identified by Foucault. Similarly, for Sahlins (1976), Western society has become a world in which the economy is mainly responsible for the production and reproduction of culture. Modern technology turns everything into a standing reserve and "drives out every other possibility of revealing" (Heidegger 1977:309). Although this second assessment exaggerates the hegemony of technology and fails to acknowledge its subversion by the simultaneous growth of other modes of disclosure such as art, literature, and religion, the first foreboding appears to have become almost a reality in late-capitalist society.[4] Even works of revolutionary art and literature have been encapsulated by the established order they criticize. Everything and everybody seems to have a price in a market economy that pulls increasingly more resources into its orbit of exchange. This conception of the world as a standing reserve is useful for an interpretive approach to the study of economy.

The economy may transform whatever turns up in its path into a standing reserve. Obviously, I do not restrict the terms economic fund and standing reserve to material resources, primary goods, investment capital, or factors of production, but include all social and cultural expressions that may contribute to the particular disclosure of the world as an economic universe. Usually, many of these manifestations are not part of the economic organization with its immediate concerns about the production, distribution, and exchange of commodities but under certain circumstances they are conceptually made available to the economic fund. These social and cultural means literally form a standing reserve to be called upon in the case of breakdown. The range of factors taken into consideration to resolve the temporary crisis is enlarged beyond the narrow concerns of the production unit or economic institution. A brief restatement of the disagreements about the choice of fishing strategies described in the last section of chapter 5 will clarify this concept of the economic fund. This restatement will be followed by a discussion of the economy as a unique mode of disclosure.

When the weather is good but not optimal, the discussions of the crew about where to fish are confined to an assessment of the amount of bait, ice, and fuel available, the phase of the moon, and the species caught by other vessels. But when the weather conditions are poor, then the strategy will be analyzed in a much wider context. What will and will not be included in this evaluation represents the economic fund that betrays the fisherman's conception of his economic universe. Boat owners will be concerned about their relation with the wholesalers who demand high-quality species of fish, high-producing boat fishermen think about the money needed to maintain their social networks, and low-producing fishermen worry about incurring greater indebtedness to the boat owners. Each group situates the fishing strategy in a unique context of personal interest and significance. The economic fund refers to whatever will be taken into consideration in evaluating the fishing strategies open to choice, even if that involves including such disparate things as a fear of supernatural wrath, a desire to appear manly, or the need to demonstrate a willingness to deliver high-quality fish to the fish dealers despite unfavorable odds. All these concerns may be entirely outside the economic realm for some but may be highly relevant for other crew members.

This interpretation of the economy as a mode of disclosure which defines everything as part of a standing reserve that people consider as economic seems tautological. However, the tautology disappears when we take the interpretations and the conflicts of the actors themselves into full account, and concentrate on how they distinguish between the economic and the noneconomic. We must pursue our understanding of the fishing economy of Camurim with those interpretations in mind and demonstrate the actor's conception of the economy and its fund of economic resources. What do the people of Camurim include and exclude from the economic fund? Where do they draw the boundaries of the economic domain? And what do they consider as the true nature of their economy?

Different conceptions of what is and is not economic prevent us from drawing a clear boundary around the economy. Economic practice in Camurim cannot be understood without transcending the social domains that lie beyond the organization of the production, distribution, exchange, and consumption of commodities. The principal economic groups of Camurim interpret economic practices differently and disagree strongly about where economic actions, resources, decisions, and responsibilities begin and end. Even the broad general definitions of economy formulated by anthropologists fall short of encompassing the complex many-faceted interpretations of the fishermen of Camurim. Let me take as an example the perceptive definition given by Godelier. Godelier (1978:55) defines the economy as

> both a domain of activities of a particular sort (production, distribution, consumption of material goods: tools, musical instruments, books, temples, etc.) and a particular aspect of all the human activities that do not strictly belong to this domain, but *the functioning of which involves the exchange and use* of material means.

Unlike many other definitions, this succinct formulation successfully includes important aspects of noneconomic domains of society. The donation to support a Candomblé center and the sacrificial offering by its spiritual leader can both be accounted for by Godelier's definition. Yet the problem with this definition is that it limits and reduces the economy to activities, relations, and exchanges, making it impossible either to incorporate the actor's interpretation of those properties into an analysis of economic practice, account for its contradictions, or include aspects that are not subject to social interaction. The myth of the economy as a bounded domain or a particular dimension of society has not been overcome. An interpretive understanding of the economy must therefore pursue the study of economic practice in all its structural contradictions, stripped of the constraints of formal definitions and enriched with the discursive conflicts of the actors. Discourse is the mode of interpretation which provides this understanding. Discourse interprets practice, while practice is an interpretation of social and cultural existence. People only become aware of practice as interpretation when the routine collapses. As action comes to a halt and people are in conflict on how to resume their social practices, the economy's alleged objectivity is demystified and its interpretive construction exposed.

If the economy cannot be defined in general terms but only as an economic mode of disclosure, then what finally is the empirical nature of that disclosure and what turns the diverse social means mentioned by the Brazilian fishermen into an economic fund? With respect to Camurim, we can say that the people conceptualize a distinction between economic and noneconomic domains of society. The economy is an ambiguous social domain whose boundaries shift according to the interpretation of the relation between economy and society. In other words, the economic mode of disclosure is defined by how that relation is worked out. The different ways in which the various social groups of the pluriform fishery interpret this relation show how the economy discloses their existence for them.

Boat owners take the position that the market is the place of contact between economy and society. Perceiving themselves to be at the mercy of a structural conglomerate of institutional market forces, they impersonalize the relations with their crews and seek to accumulate more capital, all in an effort to acquire social prestige and political power. They believe that their responsibilities to the fishermen cease with payment for labor expended. How the income is spent and what needs it can satisfy are not their concern because these issues lie beyond the economic realm. The market is an economic institution where labor and capital meet without imposing any wider responsibilities on the contractual agreement. The market with its mechanism of supply and demand is the institution that defines resources and relations as economic.

Most boat fishermen regard the market exchange of resources as only one aspect of an encompassing whole of exchange relations. They discern a fluidity in social relations which makes them believe that certain exchanges pass unhindered from one social domain to another. Their interactional conception of economy makes it hard for them to perceive a clear break between their social relations as fishermen, as members of the community, and as the providers of their families. It is exactly this particular integration of economy and society that motivates them to spend a substantial portion of their earnings in bars and brothels to cultivate close relationships with cocoa planters, ranchers, landowners, and boat owners. For boat owners, capital accumulation is strictly the result of the dynamics of economic institutions and market forces; while for the fishermen, capital goods are acquired in the sphere of banks and markets but procured in a public world of bars, drinking friendships, and favoritism. Boat fishermen, therefore, incorporate the demands of the domestic and public domain into the production unit. The fair distribution of revenue cannot be decontexualized from the noneconomic domains of society. As a consequence, the quest for patrons competes with domestic wants, even though their adherence to a dominant quasi-capitalist ideology has made them favor their socioeconomic ambitions over the needs of their families.

Canoe owners and canoe fishermen also perceive an integration of economy and society but their perception is different from that of boat fishermen. They denounce boat fishing as an expansionist mode of production, entailing the submission of both boat owners and boat fishermen to the urban fish dealers and the elite of Camurim. They remain aloof from the struggle of labor, power, and capital, precisely because they realize that they could lose control over their households if they allowed their economic independence to be sacrificed. Instead, they aspire to raise the domestic standard of living as an alternative route to social prestige. They understand that their conduct ashore and at sea simultaneously influence the success of their economic efforts and of their domestic and public lives. Canoe owners and canoe fishermen look at the economy for the reproduction of their households and production units. Everything that can contribute to the improvement of reproduction is made part of their economic fund. The different work schedules of canoe fishermen and boat fishermen are a case in point. In good weather, canoe fishermen fish on Saturdays and Sundays because the more time spent at sea, the higher the earnings will be. Boat owners and boat fishermen are also aware of the importance of good weather but they refuse to work on weekends, because that is the time they develop and reinforce their social networks. They define the weekend as public instead of economic time.

The fishermen and boat owners of Camurim take their personal interpretation of the fishing economy for granted as the truth. Yet, this truth has different referents for different social groups. Boat owners define the economy through the market, boat fishermen define it in terms of social relations, and canoe fishermen consider economic everything that contributes to the reproduction of the household. The economic actors are engaged in a politics of truth. Each social group conceives of the world with different representations; each perceives its own truth. Even though fundamental disagreements can never be resolved when people assume different perspectives of reality, they may still be negotiated or suppressed by a show of strength. Truth and falsehood are not determined by reference to objective criteria but by power relations between the contesting groups. The forced imposition of certain practices—such as the division of the catch, the authority structure aboard ship, and the credit advance system—preclude alternative ways of existence. Arguments, quarrels, disputes, sabotage, and theft are all forms of opposition to dominant practices which hint at a completely different state of affairs.

To what extent does my conception of economy invalidate the interpretations of other paradigms in economic anthropology such as formalism, substantivism, and Marxism? The problem with these paradigms is not the validity of their analyses *per se* in particular empirical instances but their claims to universality and their neglect of the constitutive force of shadow discourse. In the words of Gudeman (1986:28):

> Those who construct universal models . . . propose that within ethnographic data there exists an objectively given reality which may be captured and explained by an observer's formal model. . . . According to this perspective, a local model usually is a rationalization, mystification or ideology; at most, it only represents the underlying reality to which the observer already has privileged access.

From this self-proclaimed superior perspective, one might easily neglect shadow discourses and focus on the ideology of the landowners and boat owners which seems most forcibly to direct the local economy. If the presuppositions of the dominant ideology are in harmony with those of the universalist model, then the paradigm will fit. This coherence, however, is deceptive because the model incorporates the same ideological tenets into which the elite of Camurim has been socialized during their ascendence to a commanding position in society.

A more pluralistic approach would be to present the various interpretations of economy as competing discourses, and then ascertain which interpretation and corresponding scientific model resemble the actual economic practices more closely. This solution might be an appealing alternative to both the extreme universalism of positivist models and the threatening relativism of hermeneutics. The major drawback of this treatment is that it would reduce practice to ideology, fail to consider practice and discourse as modes of interpretation in their own right, and ignore that economic practice may simultaneously reproduce several conflicting structures. It would shut out different but structurally less articulated forms of economic practice, and assume that ultimately economic cooperation is possible only when the dominant ideology has penetrated the shadow discourses in some fundamental way. A truly interpretive study of economy must preserve discursive conflict as the dialogic process through which discourse and counter-discourse confront one another and whose negotiations may change economic practice.

Practice and discourse are two different, but dialectically related, modes of interpretation. Opposing interests among people may lead to the partial or total breakdown of practices and instigate discursive conflict and reflection. Once the conflict has been negotiated and practice takes the place of discourse, the modified practices will give rise to new experiences which will affect future discourse. The general dialectic process between discourse and practice which I have sketched here has an added significance when it functions within the dialectic of economy and society.

The vague and shifting boundary between social and economic practice, the absence of a discursive consensus on their connection, and the taken-for-granted belief in the existence of an economic domain lead to a dialectic tension between economy and society. This dialectic implies that the reproduction of economic and social practices is directed by the outcome of the discursive disagreements about the relation between economy and society. This process is not necessarily characterized by a historical trend between opposites connected through an intrinsic contradiction, but the conflicts of interest generate new practices that redefine the disputed boundary between economy and society and evoke a new context which may lead to further changes.[5] People become aware of the distinction between social and economic practices through discursive conflicts activated by recurring partial and total breakdowns, and through this awareness may try to change the relation of economy and society. The combination of the dialectic processes of practice and discourse, and economy and society accounts for the dynamic succession of changes and transformations, but also solidifications and reifications of people's social and cultural existence.

This double-edged dialectic process of transformation and reproduction becomes clearly visible in Camurim when a person switches between fishing modes or moves along the socioeconomic hierarchy. The boat fisherman who becomes a boat owner will begin to redefine the economic domain and gradually attribute different meanings to the boat and the bar because of his changing social position, power, and interests in society. This new outlook on the economy will affect his economic decisions, and thus create a new context which will further modify his assessment and interpretation of economic matters. Likewise, a fisherman who switches between fishing modes experiences interpretational shifts, which are further complicated by conflicts of action and identity. The dialectic of economy and society entails a change of social identity that bears little resemblance to the fisherman's self-perception as a free and independent producer. A former canoe fisherman who fishes on a boat does not receive the same respect from his colleagues as was common before the switch and therefore cannot claim to be independent and free, when his current position implies dependence and subjugation. Similarly, a boat fisherman who becomes a canoe fisherman cannot help spending more time at home if he wants to make a living at sea. His control over the production process will affect the relation with his former boat fishing colleagues, change his interpretation of the economy, and lead him away from the outlook assumed when he was still a boat fisherman.

The dialectic of economy and society in Camurim tends towards what I have called in the introduction, loosely consistent clusters of practice. The cultural foci represent what is important to the fishermen and give direction to their choices in life. However, the bar and the boat, and the canoe and the house, are not the driving forces behind the dialectic. The relation of economy and society cannot be reduced to the interconnected pairs of cultural foci. The cultural foci are just the centers of a referential totality of meaning. They can exist only within a world of social relationships and everyday practices.

An awareness of this referential context prevents a narrow restriction of the dialectic process of economy and society to production and consumption. Although the productive significance of boats and canoes gives paramount attention to the economic sphere, other aspects of culture are also important. Not just catch divisions, credit arrangements, and marketing strategies contribute to the formulation and interpretation of cultural foci, but also strong feelings for the family, religious and moral values, and beliefs in supernatural powers.

If the economy is dialectically related to society, and if interpretation and meaning have the importance which I claim them to have in giving direction to this process, then what explains the apparent empirical success of the management strategies of capital owners who often simply ignore the conflicting perspectives of others? The principal reason is that these strategies are in harmony with the ideology of the ruling class of Brazilian society. The actions and decisions of the capital owners of Camurim are reinforced in a national economy directed in capitalist terms. Boat owners shun many social obligations toward the families of boat fishermen under the pretext of economic rationality. They appeal to the abstract concepts of productivity, profit maximization, and supply and demand as objective guides. However, these economic concepts contain a bias in favor of the interests of capital owners whose use of this mode of analysis adds to its effectiveness. Social responsibilities can be warded off with the justification that these do not pertain to the economy.[6] Therefore, boat fishermen who want to become boat owners are forced to operate within the parameters of ruling economic concepts, even if they have alternative interpretations that expose the biases of the dominant economic model. Yet this hegemony is not all-inclusive. Boat fishermen reproduce their conception of economy in practice by procuring those interactional channels of social mobility and personal strategies of patronage which they believe to be indispensable for success. The interactional dimension of the economy will be accentuated by those actions, even though it will not succeed in superseding the institutional structure perpetuated by the boat owners and the ruling classes of Brazilian society.

The demystification of economy seems to provoke an antinomy: how can an understanding of economic behavior be attained when the economy is exposed as a myth? Has the economy not become irretrievably lost in an infinite regression of interpretations? Although the economy, its boundaries, and its institutions cannot be abstracted and understood in general terms without undue distortion, people may still define the economy as a meaningful social domain, despite their conflicts about its contextual embeddedness. Dumont (1977) and Sahlins (1976) have argued that in Western culture, the economy is the principal domain of symbolic production.[7] The economy reproduces our culture—I prefer to say that the economy discloses our culture—

through the production of meaningful commodities, and its practical reason is transposed onto other spheres of culture. Kinship, religion, art, and politics are all analyzed within the same pragmatic, utilitarian perspective. This perspective may seem very rational and matter-of-fact to us but a comparison with tribal cultures shows that other modes of disclosure may provide the principal idiom of cultural meaning. In tribal societies, social—especially kinship—relations embody the symbolic differentiation of culture. Kinship is the dominant mode from which the economy, tribal politics, and religion are constructed and interpreted. This study has shown that even though people may not challenge the economy as their principal domain of symbolic production, they may still disagree on how that mode discloses and reproduces their social and cultural existence.

My emphasis on the importance of shadow discourse does not imply that people can change their existence at will as free-floating individuals in an adventitious world of the mind. People entangle themselves, and are often forced to entangle themselves, with particular interpretations. They are, above all, social beings who through society and history influence their own existence. As has been shown many times in this book, there can be significant power differences among people that severely restrict the actions of certain groups, make them accept interpretations that offer false hopes, and allow them to realize goals at the expense of others. However, to take these social inequities and legitimizing ideologies as the true nature of human existence, or at least as a disguise of a true, hidden nature, is a self-inflicted cover-up that strengthens their mystification. The reflections of the people of Camurim on economy and society are not dispensable ornaments of the human mind. At the same time, the harsh reality of storms at sea and power struggles in public are not the only conditions of their lives. Discursive expositions intertwine and are intertwined with everyday practices. Both dimensions are interpretations of what it means to be a fisherman or a capital owner in Camurim, but one is drenched with the force of power while the other suggests different ways of being and becoming.

# Appendix

The fishermen of Camurim employ the following classification of sediment composition, depth, and sea floor reliefs to identify potential fishing spots and the fish species that reside there.

Continental Slope (*barranco, parede*)

Mud (*lama*): soft banks (*lama mole*)
firm banks (*lama dura*)

Rock (*pedra*): large exposed rocks and coral reefs (*coroa*)
high invisible rocks (*cabeço*)
large smooth rocks surrounded by mud (*lajedo*)
gravel or small rocks resting on firm clay (*burgalhão*)
shallow elevations and banks of rock or sediment (*sequeiro*)
depressions, in rocky bottoms, filled with clay (*buraco*)

The edge of the continental shelf is recognized by its rapidly declining slope. Here, large 25 to 100 kilogram fish (*peixe de fundo*) can be caught that reside at depths of 40 to 100 meters. These fish fall in the highest price categories. The species most frequently caught are sea bass (*badejo, Acanthistus brasilianus*) dolphin fish (*dourado, Coryphaena hippurus*), cherne (*Epinephelus niveatus*), *canapú* (*Promicrops itaiara*), grouper (*garoupa,*

*Epinephelus* sp.), and horse-eyed bonito (*olho-de-boi, Seriola lalandi*).

Firm soil is quite barren of fish but soft clay banks house shrimp, prawns, and mudliving species (*peixe de lama*) such as bluefish (*enchova, Pomatomus saltatrix*), ladyfish (*urbarana, Elops saurus*), various kinds of catfish (*calefate, Tachysurus luniscutsis; bagre-bandeira, Felichthys marinus; bagre, Tachysurus barbus*) and the only high quality species in these waters, sea trout (*pescada-branca, Cynoscion virescens; samucanga, Cynoscion leiarchus; pescada-amarela, Cynoscion acoupa*). Although a *buraco* is classified under the rock category, it houses the same species as are found on soft banks.

The coroa and the cabeço are the favorite fishing grounds among the classes of rocky bottoms. These large rocks and coral reefs house rockliving species (*peixe de pedra*) such as triggerfish (*peruá, Balistes* sp.), *corvina* (*Micropogon* sp., *Umbra* sp., *Pogonias* sp., *Umbrina* sp.) and several kinds of snapper (*ariocó, Lutjanus sinagris: cioba, Lutjanus analis; dentão, Lutjanus jocu;* vermelho, *Lutjanus aya*).

Aside from these three classes of fish, the fishermen distinguish two other classes. Tidal zone species migrate between fresh and salt water. The most common species are mullet (tainha, *Mugil brasiliensis*), striped mullet (*curimã, Mugil cephalus*), camurupí (*Tarpon atlanticus*), and snook (*robalo, Centropomus ensiferus*). Fishermen also catch pelagic species that are not confined to one area in particular, such as shark (*cação-martelo, Sphyrna tudes; tintureira, Galeocerdus cuvieri; galha-preta, Charcharimus limbatus;* (*cação-lixa, Nebrius cirratum*), jack (*xaréu, guariçema, Caranx* sp.), Spanish mackerel (*cavala, Scomberomorus cavalla*), and ray, (*viola, Rhinobatis percellens; arraia-amarela, Dasyatis say; arraia-manteiga, Pteroplatea* sp.; *arraia-pintada, Aetobatus narinari* (Cordell 1978; Pereira 1976).

Commercially, fish species are classified into either four or five categories. Most large fish caught at the continental slope are high-priced, while small, shallow water fish are cheaper. Thus, crews on liners fish for more profitable species than canoes and boats that exploit coastal waters. In May of 1983, the following prices per kilogram were paid by wholesalers.

| | Cr$ | $ |
|---|---|---|
| *badejo, pescada, robalo* | 500 | (1.00) |
| *ariocó, camurupí, canapú, cavala, cherne, cioba, dentão, galha-preta, garoupa, vermelho* | 350 | (0.70) |
| *cação, dourado, enchova, guarajuba, samucanga, tainha, tintureira* | 200 | (0.40) |
| *bagre, calefate, corvina, guariçema* | 150 | (0.30) |
| *arraia, bonito, peruá, viola, xaréu* | 100 | (0.20) |

This classification varies with the availability of a species. The first two categories are sometimes sold for the same price. Furthermore, these prices do not necessarily reflect local demand. Certain species such as *corvina, guariçema,* and *bagre* are much sought after in Camurim, but wholesalers do not like to buy them because they tend to spoil easily. Also, fish that have high contents of blood, such as tuna, jack, or Spanish mackerel are not easily preserved and thus fetch lower wholesale prices. Shrimp and prawns are in high demand. In December of 1982, shrimp was sold for Cr$250 ($0.50) a kilogram, and prawns for Cr$3,000 ($6) a kilogram. Wholesale prices were not available in May of 1983 because catches were very small and were sold to local consumers.

The fish prices mentioned have little significance for canoe fishermen because the value of their catches is determined through a different price-setting mechanism. For example, *samucanga, enchova,* and *corvina* make up the bulk of their catches. These fish sell for a price that is about 50 percent higher than that offered by wholesalers. Local consumers, however, cannot buy at a lower price because all fish caught by boats is transported to Vitória and Rio de Janeiro. Wholesalers would lose their sizable profit margin if they were to resell the fish locally. Hence, the separate distribution channels and different markets for the canoe and the boat fishing modes result in incompatible prices for identical goods.

# Notes

## *1. The Demystification of Economy*

1. For clarity, I am referring to the instituted embeddedness of the economy as conceptualized by Karl Polanyi and his substantivist followers who share a firm positivist belief in paradigms and universal models. For an elaboration of this positivist conviction, see note 4.
2. Bakhtin (1981:291) has expressed a similar need for hermeneutic pluralism: "Each of these 'languages' of heteroglossia requires a methodology very different from the others; each is grounded in a completely different principle for marking differences and for establishing units."
3. These thoughts have been influenced to a large extent by Gadamer (1985). Gadamer emphasizes the historicity of truth. The questions that are formulated by any interpreter are bound by the historical context in which the issues are raised. Understanding is always conditional and situated, and no scientific method can guarantee truth because understanding is always in a process of becoming.
4. Although formalist and Marxian economic anthropology have been more inclined to employ positivist methods than substantivism, all these competing schools of thought have tried to develop paradigms with a universal appeal by incorporating the latest advances in neighboring fields, such as game theory, decision theory, exchange theory, world systems theory, and ecological and cognitive anthropology. Karl Polanyi believed that "only the substantive meaning of 'economic' is capable of yielding the concepts that are required by the social sciences for an investigation of all the empirical economies of the past and present" (1957:244). He

argued that all societies can be characterized by a particular mix of three basic forms of integration: reciprocity, redistribution, and exchange. The manner in which the economy is instituted in society varies according to the form of integration that dominates the culture (Polanyi 1957:250). These positivist claims have been upheld by contemporary substantivists: "Economic anthropology is still an 'immature' field in Kuhn's (1962) sense, as there does not yet exist a general consensus among its practitioners about the core issues to be raised, the concepts which have the greatest explanatory power, and the deepest analytical conclusions about economy and society to be reached. In short, there is no prevailing paradigm. All three 'disciplinary matrices,' formalism, substantivism, and Marxism, compete for acceptance" (Dalton and Köcke 1983:35).

Orlove (1986) has made an attempt to test the validity of several approaches in economic anthropology by comparing their analyses of the coexistence of barter and cash sales in the Andean highlands. A line of research more in harmony with my hermeneutic position is pursued by Gudeman (1986). Gudeman shows that objectivist neoclassical and Marxian economic models which claim to be based on objective axioms are just as much metaphors and social constructions as the cultural models of non-Western peoples who draw more directly upon their immediate social experiences.

5. I owe my introduction to Heidegger's thought to Hubert Dreyfus. His seminar, Heidegger's Being and Time, taught at the University of California, Berkeley, in the Spring of 1986 made me aware of the close theoretical links between Heidegger and Bourdieu.
6. The postulated primacy of practice over reflection is an ontological, not an empirical, statement. Practice and discourse are dialectically related, and together constitute society and culture. Waterhouse (1981:164–170) has leveled the criticism against Heidegger that he does not perceive this dialectic because he neglects the importance of power in ontology.

Calvin Schrag (1986) has tried to reconcile practice and discourse with his concept of communicative praxis. He emphasizes the reciprocity and interdependence between discourse and action, and the importance of the hermeneutic space in which they interplay. Communication and praxis are intertwined because all communication contains speech and action, while praxis communicates messages and information. A variation on Schrag's term "communicative praxis" is Foucault's concept of discursive practice (1972). Although I am sympathetic to these terms and the explicit concern of Schrag and Foucault regarding power, I disagree with the elimination of the dialectic process of change that results from their proposed fusion of practice and discourse.

7. The notion of practice as understood by contemporary practice theorists such as Bourdieu and Giddens is very similar to Heidegger's central concept Being-in-the-World: "Each one of us is what he pursues and cares for. In everyday terms, we understand ourselves and our existence by way of the activities we pursue and the things we take care of" (Heidegger 1982:159).
8. Dreyfus (1980:7) explains the futility of the positivist pursuit for principles and models of practice eloquently: "Although practical understanding—everyday coping with things and people—involves explicit beliefs and hypotheses, these can only be meaningful in specific contexts and against a background of shared practices. And just as we can learn to swim without consciously or unconsciously acquiring a theory of swimming, we acquire these social background practices by being brought up in them, not by forming beliefs and learning rules. A specific example of such a social skill is the conversational competence involved in standing the correct distance from another member of the culture depending on whether the other person is male or female, old or young, and whether the conversation involves business, courtship, friendship, etc. More generally, and more importantly, such skills embody a whole cultural interpretation of what it means to be a human being, what a material object is, and, in general, what counts as real."
9. The West coast school of sociology embarked in the 1960s on the study of the routine grounds of everyday life. Elaborating on Alfred Schutz's notion of the epoche of the natural attitude (i.e. people take the existence of their world for granted), ethnomethodologists tried to discover the procedural assumptions and tacit rules which underlie people's commonsense knowledge by violating routine activities and conversations (Garfinkel 1967). As their studies progressed, they began to shift their research focus to the study of discursive conventions, invariant models of human perception, preconscious cognitive processing, and universals of human practice (cf. Attewell 1974; Leiter 1980; Sharrock and Anderson 1986).
10. The term "actor" is not used in this book in a dramaturgical sense as a subject who enacts social rules. Rather, actors are agents who generate social practices (Bourdieu 1987:19). Still, I have avoided the term "agent" in this text in order not to confuse it with its common usage in economic anthropology as an "intermediary" in commercial transactions.

11. My use of the term "structure" differs from that of Giddens. Giddens (1979:64–73, 1984:16–26) sees a structure as a set of rules and resources which form the structural properties of a whole of reproduced practices. Instead, I regard structure as the polysemic regularities reproduced in practice. I do not believe that unambiguous rules and norms play a role in social interaction when social practices run smoothly. They only come into being when the practices break down. A rule or norm should, therefore, be regarded as nothing more than a rhetorical leitmotiv which merely serves to delineate a vague, provisional field of discourse about practices under dispute.
12. The distinction between partial and total breakdown made in this analysis of interpretational conflicts among the capital owners and fishermen of Camurim rests on Heidegger's differentiation between a temporary and a total breakdown of practice (1962:102–107).
13. I have reserved the term ideology for the description of dominant discourse that informs hegemonic practices, following the definition of Giddens (1983:19): "I want to define ideology as the mode in which forms of signification are incorporated within systems of domination so as to sanction their continuance." I further agree with Giddens (1983:18) that "to understand this incorporation we must analyze the mode in which patterns of signification are incorporated within the medium of *day-to-day practices.*"
14. Abercrombie, Hill, and Turner also attribute a greater importance to the force relations within routine practice in maintaining the established order than to a hegemonic ideology because "pragmatic acceptance is the result of the coercive quality of everyday life and of the routines that sustain it" (1980:166).
15. Notwithstanding the importance of Foucault's study of these invisible aspects of domination, I agree with Merquior (1985:113–118) and Said (1983:178–225) that his theory of power is so broad and all-inclusive that it fails to account for radical social change. Scott's (1985) work on everyday forms of peasant resistance suffers from the same problem. Just as for Foucault everything is power, so for Scott everything seems to be an act of "capillary" resistance.
16. Nash (1979) and Taussig (1980) have written exemplary interpretive studies about how the simultaneous rejection and emulation of an ideology of inequality among Bolivian miners and Colombian peasants are translated into routine and ritual practices.
17. My concept "cultural focus" is similar to the terms "cultural exemplar" and "focal thing" used respectively by Dreyfus (1980:22) and Borgmann (1984:196). "[A] focus gathers the relations of its

context and radiates into its surroundings and informs them" (Borgmann 1984:197). Dreyfus and Borgmann use their terms to describe Heidegger's interpretation of the cultural centrality of the temple in ancient Greek society. For Heidegger, in the words of Borgmann (1984:197), "the temple not only gave a center of meaning to its world but had orienting power in the strong sense of first originating or establishing the world, of disclosing the world's essential dimensions and criteria."

18. For this reason, cultural foci are not core, primal, or focal metaphors as used by Gudeman (1986:40) because they do not idiomatically or symbolically represent practices but actively shape them. A cultural focus is neither a core, key, or epitomizing symbol as defined by Ortner (1973) and Schneider (1976). They regard these symbols as foci of cultural interest, as vehicles to communicate cultural meanings. Cultural foci do not provide cultural strategies or models for action but constitute orienting foci that shape, as well as cluster, practices.
19. I am arguing against the predominant notion in anthropology that technology is a neutral means to a cultural end. Technical instruments are designed with certain applications in mind, but the wider consequences of technology can never be foreseen. Hence, the technical implement itself will change cultural practices in unintended directions and, in effect, interpret its context in ways which may mushroom beyond people's control (cf. Ellul 1972: 103–105; Robben, in press). Giddens (1979:216–222, 1984:8–14) regards these unintended consequences of intentional acts as important homeostatic loops of social systems.
20. Zulaika (1981:34–42) describes how Spanish deep-sea fishermen have an intense longing for the family and idealize domestic life when they are fishing, but that the discrepancy between fantasy and reality at home makes them eager to return to sea. The emotional need to be close to one's family experienced at sea, ashore turns quickly into a desire for life aboard ship where social relations are more clearly defined and fishermen can display an expertise and knowledge they lack on land.
21. Interestingly enough, the most successful fishermen were usually the most inarticulate in explaining their fishing activities. I assume that their mastery had absorbed them so much in the fishing practice itself that relatively more actions had turned into habitus and that they were less often in situations of partial breakdown. In a similar vein, one can argue that informants who are socially marginal to their culture are its most eloquent interpreters and critics but not its most effective practitioners (cf. Rabinow 1977).
22. According to Dwyer, the two dominant anthropological modes—

the comparative and the relativist approach—are both guilty of objectifying the Other as an Object. These approaches deny the dialectic relation between the Subject (the anthropologist) and the Object (the informant). His assessment of these two modes applies as well to all unequal social relationships: "the activity of the Subject remains beyond scrutiny, forgoes the possibility of calling itself into question, and promotes the illusion of its own invulnerability" (Dwyer 1979:208). Dominant discourse is an expression of a native Subject who does not tolerate opposition, does not doubt its own righteousness, and treats those with less power as Objects.

## 2. *Ecology, History, and Community*

1. Devil contracts in other parts of South America have been described in detail by Nash (1979) and Taussig (1977, 1980). Taussig mentions that money earned through a devil contract by Colombian cane workers has to be spent immediately on luxury goods and cannot be invested to obtain even greater wealth. Cattle bought with evil money will die; land does not bear fruit. However, fishermen in Camurim who believe in devil contracts say that this money can be used profitably during one's entire life but that the devil will make it disappear (*sumir*) upon death.

   The local fish dealer José Periquito admitted that he believed in devil contracts but said that he did not really know how these deals were struck because he had a stronger faith in God than in the devil. "A person who does these things has only material but no spiritual wealth. This material wealth is worthless because the spirit is superior to matter. After death, a person becomes pure spirit, while his body and his wealth decay together."
2. Several historical sources mention that two Franciscan friars remained in Pôrto Seguro—located about 120 kilometers north of Camurim—between 1503 and 1505, after a shipwreck. They constructed a small church and two houses. They were attacked and eaten by Indians in 1505. Two Italian Franciscans arrived in 1515, one of whom drowned and the other returned to Italy (Pirajá da Silva in Soares de Sousa 1940:156). These friars probably visited Camurim on their trips along the southern coast of Bahia.
3. A nineteenth-century ethnohistorical account of Camurim attributes its foundation to Jesuit priests who took the original indian village under their control. "The old people said that there were only 12 houses in the settlement, inhabited by indians, and that in 1508 the first 5 Portuguese and 6 Brazilians came who brought along, as the vicar of the parish, Padre José Lopes Ferreira" (*Re-*

*vista Trimensal* 1899:485; my translation). This statement based on hearsay could not be corroborated by any of the historical resources available to me. Furthermore, the account contains several inaccuracies. Padre Ferreira could not have been a Jesuit, because Ignatius of Loyola officially established the Jesuit order in 1540. He might have been a Franciscan who visited Camurim, but the founding of a permanent settlement of colonists at this early date seems highly unlikely.

4. Pero de Magalhães de Gandavo wrote in 1576 about the colonial attempts at the genoicide of the Aimoré: "A sure means of destroying them has not been found, because they have no fixed dwelling place, and they never come out of the thicket. When we think that they are fleeing before those who pursue them, they remain behind, hidden, and shoot those who pass heedlessly by them; and in this way they kill people. All the Indians of Brazil are their enemies, and fear them greatly, because they are so treacherous a race: consequently, in a region where they are to be found, no settler goes by land to his ranch, without taking with him fifteen or twenty slaves armed with bows and arrows. These *Aimorés* are very fierce and cruel: one can not find words with which to express the cruelty of this people." (Magalhães de Gandavo 1922:140).

   Several centuries later, the Aimoré were still resisting their subjugation and continued to attack the colonists. "During the earlier frontier wars, from about 1790 to 1820, every effort was made, not merely to reduce, but to extirpate them, root and branch. Being regarded as irreclaimable savages, addicted to cannibalism and other pagan practices, and altogether no better than wild beasts, methods of warfare were adopted against them which are not usually sanctioned by civilised communities. The smallpox virus was industriously spread amongst them, and poisoned food scattered over the forests frequented by their hunters." (Keane 1884:205).

5. The French geographer Mouchez mentions both the Aimoré and the shallow river mouth of Rio Camurim as contributing to the town's lack of development, despite its rich natural resources. "The surrounding land is quite rich and produces mainly a large quantity of manioc; unfortunately the natural indolence of the inhabitants, almost all indians or mulatoes, the highly inconvenient proximity of Botocudo tribes who have come occasionally to plunder the city up till recent times, and finally the well-deserved bad reputation of the very shallow river mouth with too strong waves, are causes which prevent the development of this population" (quoted in Moreira Pinto 1899:302; my translation).

6. Only three decades later, with remarkable ecological foresight, the Queen of Portugal tried to reverse the indiscriminate exploitation of timber. In a letter to the governor of Bahia dated March 13, 1797, she wrote: "It is necessary to take all precautions to conserve the forests of the state of Brazil, and avoid their ruin and destruction." (*Revista Trimensal* 1898:19; my translation). She further declares that all forests along the litoral and river margins will become royal property. Former coastal concessions will be revoked and the owners will be compensated with land in the interior. Although the burning and cutting of royal forests called for severe punishment, the decree did not seem to have had much impact. Still, Filho (1976:44) states that similar decrees obliged colonists to explore the hinterland, leading eventually to the establishment of cocoa plantations in Bahia.
7. Demographic figures of Camurim from 1850 till 1980. Sources: *Censo Demográfico* 1900, 1920, 1940, 1950, 1960, 1970, 1980; Moreira Pinto 1899:302; *Revista do Instituto Geographico* 1902:102; *Revista Trimensal* 1898:23; 1899:485

| *Year* | *County* | *City* |
|---|---|---|
| 1850 | | 350 |
| 1863 | 2,000 | |
| 1900 | 5,514 | 1,400 |
| 1910 | 7,228 | |
| 1920 | 11,445 | 2,998 |
| 1930 | | * |
| 1940 | 16,623 | 1,410 |
| 1950 | 33,104 | 1,589 |
| 1960 | 61,164 | 2,200† |
| 1970 | 31,591 | 3,685 |
| 1980 | 26,443 | 5,708 |

*The demographic census was not conducted in 1930.
†The county of Camurim was divided in two after the 1960 census had been conducted and the new county of Serrania was founded.

8. The noun *desbravador* is a derivation of the verb *desbravar* which means to tame, to domesticate, to reclaim, to cultivate, and also to settle.
9. Poole (1981:215) lists a number of forces that might induce men to either accept or reject employment at sea. Incentives to go to sea relevant to this study are: an early occupational choice, satisfactory earnings, few economic opportunities ashore, and family problems.
10. A *lajota* is a clay brick with hollow cells that enhance insulation and make these bricks three times larger than a solid handmade

brick. Although solid bricks are stronger and more durable, many people prefer hollow bricks because they require less mortar and labor during construction. Furthermore, the price of one *lajota* buys only two-and-a-half solid bricks.

11. Tourism offers very little other employment to men but it allows some women to earn cash money. They work in the homes of the summer residents and sometimes develop clientelistic relationships with their employers.

 Camurim has barely any permanent employment for women. Most school teachers are women, and about fifteen married women work as charwomen in the hospital, the city hall, and in the schools. The only work available to poor, unschooled women is brickmaking, streetcleaning, washing and sewing clothes, cleaning shellfish, and doing domestic work for the town's elite (Robben 1988a).

12. The two beach seines fishing along Camurim's beaches are about 70 meters long, 5 meters wide at the sleeves, and 10 meters wide at the center. Nets are never bought new but rather restored piecemeal by each new owner. The two current owners, who also own motorboats, purchased the beach seines when fishing in the tidal zone was more profitable and a dozen or more nets adorned the beaches of Camurim. Overfishing in the shallow coastal waters by canoe fishermen and shrimp trawlers has substantially diminished the catch and has obliged owner after owner to abandon his beach seines.

13. Pálsson and Durrenberger (1983) have challenged the "skipper effect" in their detailed studies of Icelandic fishermen. They argue that the so-called expertise and intuitive qualities of the captain are a myth sustained to enhance their prestige. The continued success of captains, so they argue, depends—in the long run—on the size of their boats, the number of fishing trips, and luck.

14. Unlike in other regions of Northeast Brazil where the continental slope is considerably closer to shore, medium-size boats of 8 to 9 meters cannot alternate between fishing in coastal waters and at the edge of the continental shelf. The storage hold is too small and the voyage too long to make such fishing trips profitable.

15. Souza (1976) gives a penetrating account of the dangerous handlining expeditions hundreds of miles offshore. The mother ship launches at daybreak up to thirty small row boats over a distance of several kilometers, each manned by one fisherman, and gathers the men and their catch at sunset. Some fishermen from Camurim have fished on these liners from Vitória, and one told me a harrowing story of being lost at sea for three days before the mother vessel recovered him.

16. The Miskito Indians of Eastern Nicaragua gave a similar explanation to the rapidly declining catches of green sea turtles when they began to sell turtle meat to export companies between 1969 and 1977. The Indians could not understand that the turtles they had been hunting for centuries were suddenly endangered. They believed that the Turtle Mother had moved the turtles to Costa Rica out of anger for their commercial exploitation. Hence, the Miskito saw no reason to exercise any restraint in turtle fishing because the animals were as abundant as they had always been, only now they were harder to catch (Nietschmann 1973:199–202, 1979:60).

## *3. Canoe Fishing Along the Atlantic Coast*

1. Canoe owners and canoe fishermen are in many respects so similar that I will refer to them most often as canoe fishermen. I will mention both groups separately when the distinction is relevant.
2. Giddens explains that socialization is not merely a phase of human development but a lifelong process because "socialisation does not just stop at some particular point in the life of the individual, when he or she becomes a mature member of society. That socialisation is confined to childhood, or to childhood plus adolescence, is an explicit or implicit assumption of a good number of those who have made use of the term. Bot socialisation should really be understood as referring to the whole life-cycle of the individual" (1979:129). In a similar vein, Dundes (1983:238) indicates that people do not have one identity, but have many personal and social identities, and that they acquire new ones and drop old ones throughout their lives.
3. Giddens (1979:121) correctly criticizes Mead for conceptualizing the self as a harmonious relation between the I and the Me, making it difficult to account for identity crises. Chapter 6 will address the conflict between personal and social identity among fishermen who switch between fishing modes while holding on to the social identity of their former peers.
4. Some form of territoriality might have existed in the 1960s, when the number of beach seines increased due to a rising demand for fish. Since only a limited number of beach seines can operate at the same time on a particular stretch, there must have been some system of rotation. Old fishermen have told me about conflicts between net owners but the information is sparse. Alexander (1977) describes the consequences of a growing number of

beach seines on territoriality in Sri Lanka. His case might be comparable to the situation in Camurim before the arrival of boats.

5. Acheson (1975:189–191) makes a distinction between nucleated and perimeter-defended fishing areas among the New England lobster fishermen. Nucleated areas are located close to the harbor of embarkation and are fished in only by one group. Residence in the harbor is usually sufficient to gain access to the waters. Territorial claims are more intense in perimeter-defended areas farther offshore that are exploited by groups from several ports. Intrusions meet with much greater opposition and entry is more limited. In these areas, kinship becomes the most important criterion for access. The opposite is the case in Camurim. Areas close to shore are more strongly defended, while distant waters are difficult to claim when fishermen practice handlining.
6. The members of a corporate group may share ritual kinship relations but these ties do not have any significance in recruiting new members because god- and co-parenthood ties cut across neighborhoods, generations, and social classes.
7. Residence among Brazilian fishermen is neolocal but a new household is preferably established next to the parental house of either bride or groom (cf. Forman 1970:108). A recent housing shortage has made it difficult to fulfill this postmarital residence practice.
8. *Angelim* (*Andira nítida*), *barriga-d'água* (*Hydrogaster trinerve*), and *oiticica-da-mata* (*Clarisia racemosa*) are the kinds of trees most often used for canoes. A suitable tree must have a circumference of at least 0.3 to 0.35 meters to make a canoe that is three to three and one-half *palmos* (handbreadths) wide (0.66–0.77 meters) and twenty to twenty-five *palmos* long (4.4–5.5 meters).
9. Attempts to motorize canoes with outboard engines have been unsuccessful. The danger of loss made them unsuitable in bad weather, the cost of fuel and maintenance were high, and the rise in productivity proved too small.
10. Nets are made of 0.3, 0.6, and 0.7 millimeter monofilament nylon line. Each diameter corresponds to a different mesh size. Nets with fine mesh made of 0.3 millimeter (*tainheira*) are more fragile and entangle small fish but catch larger quantities than the 0.6 and 0.7 millimeter wide-mesh nets (*caçoeira*).
11. Compare the empirical situation of netmending in Camurim with the following observation about the relation between people and objects in modern society: "The glance that is cast upon a technical object—passive, concerned only with the way it works, with its structure, how it can be taken apart and put together, fas-

cinated by this backgroundless display all in transparent surface—this glance is the prototype of a social act." (Lefebrve 1971:49)

12. Verbal dueling is a common characteristic of these unfocused social gatherings which will not be discussed here. Especially when discursive conflicts have reached a point where a further discussion is fruitless, one fisherman will try to outsmart the others with the use of proverbs and common sayings. Joking becomes a way to transcend the disagreement (cf. Bricker 1976; Dundes, Leach, and Özkök 1972; Lavie 1984).
13. Bombeiro is such an appropriate term to describe these features, that a professional fisherman who has not gone fishing for several days calls himself mockingly a bombeiro.
14. There is a clear parallel between the situation of the canoe fishermen of Camurim and both the native American fur trappers and Newfoundland fishermen before the start of the twentieth century: "Both native American peoples and Newfoundland fishermen 'sold' the product of their own labour, rather than their labour-power itself. . . . Both were dominated at the point of exchange, rather than production, and although the severity of the domination at the point of exchange permeated all other aspects of their social life, still in the work process itself and in the social relationships within which the work process was organized (family and village life) they retained a certain autonomy." (Sider 1980:15).

## 4. *Boats on the High Seas*

1. Elmício gave a very different account: "When the Navy lieutenant came, there was trouble because Francisco was without fishing documents. I couldn't even touch the boat, whose bottom was rotting in the river. Fortunately, he let me repaint the hull and, after seven months, the boat was fishing again. The President of the Fishermen's Guild forged all the right papers to give the late Francisco his documents and the case was closed."
2. In May 1983, a boat owner borrowed Cr$1,100,000 ($2,200) to buy a new engine. The loan was to be paid off in five years. The first half-year was a grace period during which he did not pay any installments. After this period, he paid one installment each month and interest on the remaining debt twice a year. The amount was apportioned into 54 monthly installments. His first monthly payment of Cr$10,000 was on December 30, 1983, the second on January 30, 1984, and so on till the sixth payment of

Cr$10,000 on May 30, 1984. The seventh payment was increased to Cr$14,000 and paid on June 30, 1984, and so on. The loan officer gave me the following breakdown of installments:

| *Number of months* | | *Amount of installment* | *Period of payment* |
|---|---|---|---|
| *6* | × | *Cr$10,000* | *12/30/1983–5/30/1984* |
| *12* | × | *14,000* | *6/30/1984–5/30/1985* |
| *12* | × | *17,000* | *6/30/1985–5/30/1986* |
| *12* | × | *25,000* | *6/30/1986–5/30/1987* |
| *11* | × | *30,000* | *6/30/1987–5/30/1988* |
| *1* | × | *38,000* | *5/30/1988* |

In addition, interest payments had to be made on June 30 and December 31 of each year. Clearly, this was a very beneficial arrangement, because on December 29, 1983, the rate of exchange for U.S. dollar was already Cr$980, thus reducing the entire debt to only $1,125 before any payment had been made. Unfortunately, the high inflation rate has also reduced the purchasing power of new borrowers. The maximum amount available under these terms in May of 1983 was $2,250. The maximum amount—in 1980—bought a new 8-meter motorboat, but the 1983 amount bought only an engine or a medium-size 8 to 9 meter sailboat.

3. The species of sardine (*sardinha-cascuda, Herengula clupeola*) in the tropical waters around Camurim is different from the species (*sardinha-verdadeira, Sardinella aurita*) of the more temperate waters of southern Brazil. The boat fishermen claim that the southern quality is better bait because of its higher content of oil, thus attracting more fish. However, canoe fishermen who occasionally fish with the local species argue that fresh bait is always better than the salted sardines from the south. Irrespective of whoever is right, there are simply not enough local sardines available for boat fishermen to make this choice.
4. A known practice is to secretly sell high quality species to one dealer and hand the remainder to the other dealer. The boat owner deprives the second dealer of a profitable retail commission and breaks the oral agreement that, in exchange for giving supplies on credit, the middleman has the right to buy the entire catch at preestablished prices. The illicit transactions are either done ashore or at sea. The crew moors the boat before reaching the harbor, unloads the fish and, as it is called, "spawns ashore" (*desova na terra*). A less common way is to "lay eggs at sea"

(*desova no mar*) and transfer part of the catch to small boats that are temporarily without a dealer. Although these improper dealings are made by some owners, most of them prefer not to jeopardize their good standing with intermediaries on whom they are so dependent.

5. Brazilians traditionally eat fish during Holy Week. The prices rise rapidly because of a shortage in the cities. Many boats in Camurim switch during the fortnight before Easter to salting fish so that they can load the storage hold with as much fish as they can catch, of whatever quality. Although fishermen dislike salting fish immensely, they look forward to this period as a time when they can pay off their outstanding debts. Fish that is usually thrown back into the sea, or is given to some poor relative, is salted and sold for a good price.
6. The expense for fuel for a roundtrip between Camurim and Rio de Janeiro is about Cr$50,000 ($100), the cost of ice is Cr$60,000 ($120). Finally, the dealer has to pay export taxes of Cr$20 ($0.04) per kilogram of shrimp and Cr$10 ($0.02) per kilogram of fish transported from the state of Bahia to the south.
7. The depreciation of the truck is difficult to ascertain because of the bad roads and truck service to other communities along the coast. The wage of the truck driver is negligible in the overall costs, because he does not earn more than Cr$40,000 ($80) a month.
8. Acheson (1985) interprets a similar mistrust of lobster dealers in Maine as a lack of understanding of the marketing processes among fishermen. Although dealers withhold market information, Acheson argues, their large number prevents widespread price fixing because the producers can always find a better-paying outlet of lobster. In Camurim, however, the fish prices do not fluctuate directly with supply and demand, and there are seldom more than two wholesalers operating in the region.
9. The boat owners and fish dealers are very secretive about this commission. The boat owners claim that they have the overhead cost of constructing or renting a space near the waterfront to keep equipment, fuel, and boxes of ice to store fish in case the trucks break down or the roads become impassable. They also weigh the catch in these places and are commonly accused of meddling with the scales. However, the rent is seldom more than Cr$3,000 ($10) a month, while the retail commission on large catches almost doubles the boat owner's income.
10. I have found the comprehensive and well-illustrated book of Andres von Brandt (1972) to be the most useful guide on fishing techniques for anthropologists because it discusses the entire range

from simple gathering and handlining methods to the most advanced long-lining and trawling techniques.

11. Some large liners have also been experimenting with trawl nets but the results have been disappointing. Their liners' size and stronger engines do not lead to proportionally larger catches, demonstrating once again that large boats cannot operate as profitably in shallow waters as small boats can. The shrimp banks of Camurim are relatively small and not as rich as those along the coasts of Espírito Santo and Rio de Janeiro, where large trawlers operate successfully.
12. The number of gill nets in Camurim presages overfishing. The canoe fishermen own, all together, about 200 nets and the boat owners have roughly 800 nets. The total length of these nets is 100 kilometers, putting a very heavy strain on the coastal habitat. Boat owners complain that when they began netfishing, 10 nets used to catch as much fish as 40 nets do at the present time. As a result, the investment costs have risen significantly: 35 nets cost Cr$1,500,000 ($3,000), or the purchase price of a small motorboat. Almost thirty boat fishermen also own nets but only a few men have more than two or three gill nets.
13. The characterization of the boat as a social institution is, of course, relative. A too rigid use of the term has come under a justified attack in maritime studies. Compared to social institutions such as prisons and mental hospitals, the isolation of the fishermen is quite short, the discipline is less strict, there are possibilities of moving up in rank, and there exist more institutional channels for grievances (Poole 1981:209–210). Zulaika (1981:x) points out justifiably that the wider cultural context significantly shapes the behavior on ships despite the social isolation from the life ashore.
14. Of the literature on the sea that I am familiar with, I found Peter Matthiessen's wonderful description (1975) of the experiences of a crew of turtle fishermen from the Cayman Islands to come closest to the conversational style common among the boat fishermen of Camurim.
15. Some low-producing fishermen are said to be plain unlucky but their misfortune is believed to end if they decide to work hard. Poor luck is often attributed to a colleague's evil eye or a spell. Good luck is bestowed upon a person by God or by the Sea Goddess, especially when that person is in dire need but has faith in these deities. Zulaika (1981:65–94) has made a thorough analysis of the importance of luck among Spanish deep-sea fishermen. He concludes that there is no direct relationship between effort and output and that, therefore, luck is regarded as a form of arbitrary causation. Probably, luck does not have a sim-

ilar explanatory prominence among the boat fishermen of Camurim because handlining is a fishing method through which individual differences in work performance are immediately visible. The extended cod-trawling expeditions of Spanish fishermen, instead, are collective undertakings in which calling attention to personal ability would be disruptive.

16. There is another interesting theme about the value of economic goods running through these explanations about the high price of shark fins which, unfortunately, can only be hinted at here. In their understanding of the economy, many Camurimenses assume implicitly some degree of compatability between the use value and exchange value of a product. Snook is not only expensive because it is in high demand, but because it is the most delicious fish available. So, because shark fins do not have any use value in Camurim, they must have some unknown substance which can explain their high exchange value. Drugs, nuclear energy, radioactive material, and computers are examples of such valuable goods that speak to the minds of the people of Camurim.
17. There is, of course, not a perfect correlation between the interpretations of the practice of boat fishing and the economic position of the men. Some boat owners and high-producing fishermen are content with the status they have achieved, and many low-income fishermen still adhere, at least in the presence of boat owners, to the interpretations of high-income colleagues, although their earnings do not justify such a positive outlook. These poor fishermen regard their low earnings as nothing more than a temporary lapse in performance, and they resent boat owners and high-producing fishermen for classifying them as low-producing fishermen.
18. The different approach to risk between canoe fishermen and boat fishermen brings immediately to mind the debate about the rational versus the moral nature of peasants. One might reason that the canoe fishermen adhere to the principle of safety first and do not want to maximize their income at the risk of their subsistence and give more importance to the welfare of the households of the corporate group (cf. Scott 1976). Boat fishermen seem to act more as rational peasants, who look first after their individual interests and are disposed to risk their surplus production if the economic circumstances promise to yield a considerable gain (cf. Popkin 1979). Yet I hesitate to couch the differences between canoe fishermen and boat fishermen in these terms because of the complex changes these men undergo when they switch from one fishing mode to another. Their interpretation of economic practice changes so dramatically, as will be

seen in chapter 5, that such a dichotomy seems to do too much violence to the ethnographic data.

## 5. Boats and Canoes in Coexistence

1. In conversations with fishermen, I found that all groups justified their preferences in the same rhetorical way. They would extoll the virtues of their own fishing mode and only mention the negative aspects of the other mode in a very stereotyical, almost caricatural, manner. I would usually ask the informant to substantiate his generalizations by his personal experience in order to understand the basis of this negative assessment.
2. Erikson (1959:122–124) also uses the term "breakdown" in his studies on identity, but he refers mainly to a phase of psychological confusion in late adolescence during which a young person has to choose between the many new roles that are suddenly available. Still, his description of this dilemma applies effectively to switching fishermen torn by opposing loyalties, "A state of paralysis may ensue, the mechanisms of which appear to be devised to maintain a state of minimal actual choice and commitment with a maximum inner conviction of still being the chooser" (1959:124). Especially those fishermen who have switched to boats assure their worried former colleagues that they are in complete control, that they will not allow themselves to run up debts, and that they can return to canoe fishing whenever they want to. In some cases, the fishermen were deluding themselves about this control, and were slipping deeper and deeper into debt.
3. The close relation between the start of the life cycle and the choice among fishing modes was suggested to me by Raymond Kelly. The development of a personal identity is strongly influenced by childhood experiences. A boy who grows up in the harbor quarter and seldom sees canoe fishermen in action will most likely identify with the boat fishermen in his neighborhood and find this self-image reaffirmed throughout life. These differences in the formation of the personal identity of future canoe fishermen and boat fishermen, rather than an objective weighing of the pros and cons of the two fishing modes, significantly guides their choice.
4. Federal labor laws require boat owners to have written work contracts with their crews, and to pay for their insurance costs and dues to the Fishermen's Guild. These regulations are usually ignored and thus allow canoe fishermen to switch freely between boats and canoes.
5. Many boat and canoe fishermen believe that the president fills

his pockets with the Guild's funds and they refuse, therefore, to pay the Cr$30 ($0.60) monthly dues. Boat owners are supposed to pay Cr$1,000 ($2) a month per boat and a 5 percent sales tax on their fish sales, but only a few men pay their share. This general refusal to support the Guild leaves many men without insurance and gives retired fishermen only small pensions.

6. I do not have any hard data to support the claim by boat fishermen that their divorce rate is higher than that among canoe fishermen, but my informal queries pointed in the same direction. The difference was largely attributed to the greater incidence of extramarital affairs, or at least the suspicion thereof. Azevedo (1965:305) mentions that adultery on the part of the woman and their ill treatment, or the abandonment of the home, by men seem to be the most common causes of divorce.
7. I have only canoe fishing data for this period in early summer, because I had not yet established enough rapport with boat owners to obtain accurate catch figures.
8. The importance of waiting is clearly expressed in the common saying: "A esperança está na água" (hope resides at sea). The noun *esperança* derives from the verb *esperar* which means "to hope" as well as "to wait." Boat owners manipulate this double meaning discursively and tell the fishermen, "Quem espera, sempre alcança" (who waits, always succeeds). Boats and their crews have to be exposed to the opportunities at sea to produce the desired yields; they have to "wait" at the best fishing grounds and let nature run its course. The fishermen usually reply in these discursive confrontations that "Quem espera, desespera" (who waits, despairs). The boat will not catch any fish for them, but they have to make the effort to procure the fish at sea.
9. There is a growing literature in decision making analysis which deals explicitly with fishing strategies. Davenport's (1960) application of Von Neumann and Morgenstern's game theory is the most well-known study. The major problem with the use of rationality and information processing models is that they treat decision making as a virtually context-free sequence of a limited number of individual choices in isolation from other domains of culture. Ortiz (1983) has shown the dangers of this narrow approach by pointing at the complex social context of management strategies.

## 6. *Sea, House, and Street*

1. I soon heard of many other incidents of atraso. Atraso may cause a person to fall ill, a fisherman to have bad luck, or a house to

catch fire. *Atraso* may be caused by others or may be self-inflicted. The harmful glance of a person with an evil eye may lead to *atraso*, but *atraso* can also occur when customary ways are broken. A fisherman who comments that "Minha pescaria está atrasado" (my fishery is lagging behind), means to say that he is catching fewer fish than usual, that someone is bewitching him or, finally, that he, himself, has done something wrong. In a similar vein, poor boat fishermen often say that "Os donos gostam de botar a gente para atrás" (the boat owners like to put us at a disadvantage). They cannot get ahead in life because the boat owners prevent them from being more successful.

2. This position is supported by proxemic studies which clearly demonstrate that people are generally unaware of their spatial behavior. These proxemic conventions are not taught but assimilated through imitation (Watson 1970:90).
3. Other spatial structures that constitute habitus are churches, schools, hospitals, gardens, and prisons. A good example is Bentham's Panopticon, an eighteenth century design of a model prison intended to instill discipline through the assumed presence of a guard who could not be seen by the convicts. As Foucault (1977:201) explains, the Panopticon was designed in such a way "that this architectural apparatus should be a machine for creating and sustaining a power relation independent of the person who exercises it; in short, that the inmates should be caught up in a power situation of which they are themselves the bearers."
4. Oliver also emphasizes the cross-cultural importance of the house in people's lives. "But the dwelling is more than the materials from which it is made, the labour that has gone into its construction, or the time and money that may have been expended on it: the dwelling is the theatre of our lives, where the major dramas of birth and death, of procreation and recreation, of labour and of being in labour are played out and in which a succession of scenes of daily lives is perpetually enacted" (1987:15).
5. Through a letter from the local parish priest, I learned that the sound system was installed in 1985. He gave the impression that the more wealthy townspeople, especially, were not very happy with the popular music invading their homes.
6. This neighborhood is also the location of the Candomblé centers in Camurim. It is a clear expression of the difference in the sociospatial and spiritual prestige between the Roman Catholic Church—which represents the official religion of the elite and has been constructed in the center of town—and Candomblé which has more popular support but which is considered evil by many influential townspeople and therefore would never be tolerated in a more prominent location.

7. The only occasion during which I heard people comment explicitly on the sociospatial organization of Camurim was when a fisherman would instruct an inexperienced teenage fish peddler on how to go about selling his fish. He would explain which neighborhoods to avoid because the residents did not like certain species of fish, would only buy a more expensive kind, or spent too much time haggling for a lower price.
8. Barrington Moore (1984) discusses the cross-cultural pervasiveness of the public and the private, and traces our Western notions to ancient Greek and Hebrew society. Hannah Arendt (1958), in turn, emphasizes the difficulties in keeping the two domains apart because one is defined through the other and the boundary between the domains has shifted constantly in the history of Western civilization.
9. In a more recent publication, DaMatta (1987:51–69, 164–173) refers to the house, the street, and the supernatural as the three principal spheres of Brazilian society. In this study, I have discussed this important supernatural dimension only insofar as it related directly to economic practice.
10. The failure to make a clear distinction between work and labor lies at the root of the common confusion of the public with the economic domain. Arendt (1958) has argued that "labor is the activity which corresponds to the biological process of the human body" (1958:7). Labor produces "means of subsistence and reproduction of labor power" (1958:143), while "work provides an 'artificial' world of things" (1958:7) which has a "worldly permanence" (1958:143). In other words, labor perpetuates society, while work transforms labor power into culture; work reifies cultural meanings into public objects. The indiscriminate use of "labor" and "work" as synonyms has caused the confusion of the public quality of work with the social quality of labor.
11. People in Camurim only greet each other when they enter or leave a house. At their departure, the host invariably says to the guest, "Está cedo ainda" (it's still early) or "Fica mais um pouquinho" (stay a bit longer). He may also invite the person for dinner, an offer that is to be declined politely (cf. Pierson 1951:121). However, fishermen will never exchange greetings when they enter a bar, get aboard ship, or join a group of men at a street corner, on the beach, or in the harbor. A greeting would be appropriate only if the person had been away for some time.
12. Lewcock and Brans (1977) suggest that in maritime cultures there is often a correspondence between the sociospatial organization of boats and houses. "Symbolic relationships between sea and land, man and community, the individual and his ancestors, death and the perpetuation of life and so on, are given expression in

buildings, meeting places, and even in the layouts of villages" (1977:116). Several boat fishermen in Camurim referred occasionally to the boat as a house and at other times drew an analogy with the human body. Unfortunately, I cannot explore these relations here.

13. For additional historical analyses of the spatial organization of the eighteenth century European house, see Collomp (1986:509–519), Elias (1978:160–168), Guerrand (1987:332–335, 349–356), and Ranum (1986:219–232).
14. For a description of the construction and geographical distribution of wattle and daub houses in Northeast Brazil, see Freyre (1937). Pierson (1951:42–47) also provides a detailed description of the material culture of these peasant dwellings of Northeast Brazil.
15. The transformation of wooden into brick houses passes through several stages in which increasingly better construction materials are used and rooms are enlarged and added. A brick façade is the first improvement to a wooden house. Next, the wooden structure is replaced with a construction of posts of reinforced concrete, and walls are made of hollow bricks (*lajotas*). A bathroom is the next addition, followed by the construction of extra rooms and a front porch. Once the house has reached the boundaries of its site, the owner might decide to sell it and build a new house on a larger lot and at a more prestigious location. That house will be made of solid bricks and be covered by more expensive roof tiles.

    A comparison between houses owned by people in Camurim and shantydwellers in Rio de Janeiro shows that the wooden houses of poor boat fishermen measure up to the houses owned by squatters in the higher echelons of the slum neighborhood. Their houses correspond to the eleventh of the fourteen architectural stages distinguished by Drummond (1981) in his study of a large squatter settlement in Rio de Janeiro. The houses of the boat owners resemble the final stage. The significance of this comparison is that the relatively decent housing in Camurim is mentioned by fishermen who have worked in Vitória, Rio de Janeiro, and São Paulo as an important impediment on cityward migration.
16. The work on proxemics by Edward Hall has demonstrated convincingly the interrelation of social interaction and the organization of space: "Fixed-feature space is one of the basic ways of organizing the activities of individuals and groups. It includes material manifestations as well as the hidden, internalized designs that govern behavior as man moves about on this earth . . . Even the inside of the Western house is organized

spatially. Not only are there special rooms for social functions—food preparation, eating, entertaining and socializing, rest, recuperation, and procreation—but for sanitation as well." (Hall 1969:103)

Paul-Lévy and Segaud (1983) have provided an excellent compendium on the study of space.

17. The *platibanda* was introduced to Brazil in 1927 by the European architect, Gregori Warchavchik. His private home in São Paulo with its symmetrical, rectangular façade became the first example of modern Brazilian architecture in which function directed form (Chazan 1977:38–39). The *platibanda* has become the most important contemporary expression of rural vernacular architecture in Brazil.
18. *Pratibandas* have no other functions than being front walls. In fact, *pratibandas* are architecturally inferior to the façades of the traditional chalet-style houses with overhanging roofs. The strong winter winds do not affect the façade, but rain that hits the back of the elevation seeps into the houses and forms small pools of water in the front room.
19. The porch is not yet a significant marker among canoe fishermen because the cost of construction is too high for most households. Only one canoe owner has a front porch but it is so small that it barely holds two chairs. The owner demolished the façade and the front room to make place for the porch, and extended the house towards the back.
20. Canoe owners and canoe fishermen ascending on the ladder of prestige will pursue their material wishes in roughly the following order: *pratibanda*, gas stove, couch and armchairs, cupboard, brick walls, veranda, refrigerator, formica table and chairs, radio, bicycle, and television set.

## 7. *Social Relationships and Family Life*

1. As Alfred Schutz has written in another context about the separation of a man from his family: "To him life at home is no longer accessible in immediacy. He has stepped, so to speak, into another social dimension not covered by the system of coordinates used as the scheme of reference for life at home. No longer does he experience as a participant in a vivid present the many we-relations which form the texture of the home group. His leaving home has replaced these vivid experiences with memories, and these memories preserve merely what home life meant up to the moment he left it behind. The ongoing development has come to a standstill" (Schutz 1976:111–112).

2. Stallybrass and White (1986) go two steps beyond the relation of space and practice implied here by including the human body and psyche in their analysis. They claim to have identified a basic process of classification between fundamental oppositions such as high and low, polite and vulgar, and clean and dirty, through which the European middle class can identify itself with that which it abhors but at the same time desires. "The high/low opposition in each of our four symbolic domains—psychic forms, the human body, geographical space and social order—is a fundamental basis to mechanisms of ordering and sense-making in European cultures. Divisions and discriminations in one domain are continually structured, legitimated and dissolved by reference to the vertical symbolic hierarchy which operates in the other three domains" (1986:3).

   For another approach to the relation between house and body, see Bachelard (1969). In this book, I cannot explore these interesting connections, but the many tabus and proverbs used by the fishermen which refer explicitly to the body suggest a fruitful line of further research. Ribeiro (n.d.: 103–108) gives an extensive list of such proverbs common in Brazil.
3. I have translated the expression "Vai tomar no cú" as "don't bother me," but literally it means, "take it up your ass," or "go and commit sodomy." The expression is used in everyday speech among strangers as well as couples, but the strength of the expression varies with the social context. In this case, Zé Peroba simply did not want to be bothered, but his recent marriage and the allusions in the conversation to the wedding night added an underlying sexual meaning to the exclamation.
4. Aside from love (*amor*), there is passion (paixão), which refers more to the physical attraction of a couple and is regarded as necessary for an amorous relationship between man and woman.
5. Their dedication to the household does not simply imply that canoe fishermen will refrain from brief sexual affairs. In summer, some men will make advances to female tourists whom they consider to be outside their community and hence harmless to their domestic life. Yet, even though their occupation makes adultery by canoe fishermen less common than by boat fishermen, the ideology of male superiority (machismo) makes certain men desire a mistress. But again, the incidence of such enduring extramarital affairs among canoe fishermen is relatively small. Occasionally, a man may visit a prostitute, but he never visits brothels as often as boat owners and boat fishermen.
6. It is significant that the Portuguese word for eating (*comer*) is also the vulgar term for sexual intercourse. See DaMatta (1986:58–64) for an analysis of the double meaning of the term *comer*.

7. The liberty of boat owners and boat fishermen does not imply that they can neglect the sexual needs of their wives. Although the men might still spend the evening in a brothel, on the day that they return from sea they will always pass a couple of hours with their wives "to leave a credit advance at home" (*deixar o vale em casa*), as they euphemistically call sexual intercourse immediately after a fishing trip.
8. Although the male head of household is the principal breadwinner and his wife keeps her income for herself, other members of the nuclear family may also earn money. Sons of professional fishermen are generally also involved in fishing. Their income, however, is not added to the household budget from which domestic expenses are paid. Adolescent sons use their irregular earnings to hang out in nightclubs, visit prostitutes, or gamble with playing cards. They only contribute fish to the household. Some occasionally work as construction workers, but only two young men have become masons. Fishermen would like to see their sons become carpenters, masons, and electricians, but few are able to break through the customary passing of trades on from father to son. They cannot easily find apprenticeships and will usually take up fishing after finishing primary school. When they reach twenty years of age and are close to getting married, they begin to work full time, save money to buy fishing nets, and prepare themselves financially for an independent household.
9. One captain gave up his position on a successful boat to be near his wife, who was hospitalized in the city of Vitória. He worked on large fishing vessels and spared neither cost nor effort to have his wife attended to by the best physicians. One boat owner sold one of his two boats and a large number of nets to get expert treatment for his autistic daughter. Whenever he is ashore, he tenderly carries the girl to the river, where she sits closely beside him as he mends his nets.
10. In a survey of the fishing village of Arembepe in northern Bahia, Figueiredo (1983:98) established the number of households with a woman as the chief economic provider at 29.5 percent. She emphasizes that this high number of matricentric households is not a contingent, but a structural, phenomenon. I do not have any comparable statistics from Camurim, but if one were to add the households of low-producing fishermen to those of many salaried workers, I would suspect that around 20 percent of the heads of households are women.

## *8. Public Image and Social Prestige*

1. Stores are, of course, also public places but they have more in common with workplaces in which people enter contractual arrangements. Churches, in turn, are public spaces specifically designated for ritual purposes, while government offices are within the political and bureaucratic realm.
2. Participants who prefer liquor to beer may of course drink more slowly than beer drinkers, but rounds usually consist of only one kind of beverage. The consumption of the same kind of drink enhances the solidarity of the men. Being only a moderate drinker myself, I was singled out repeatedly with comments such as, "Hey, look, Antônio is not drinking at all!"
3. I will not discuss nightclubs (*boate*) and dance-halls (discoteca) because these establishments attract mainly teenagers and young adults who are not married.
4. "It is against this background of human association with its dual potential for individual adjustment and social solidarity that we must view the strong pressures on seamen to drink. Drinking is a central and approved activity among seamen. To refuse to drink with shipmates may be viewed seriously" (Nolan 1976:81).
5. Only one canoe owner has remained childless after two marriages and a score of concubinages. People believe him to be sterile, but at the same time they tell numerous anecdotes of his extraordinary physical power. This man is compensating with strength what he lacks in fertility and therefore still receives the respect proper to a man.
6. The manifestations of masculinity are interactional expressions of machismo, an attitude of male superiority that makes men act with arrogance, aggression, intransigence, and self-importance (cf. Brandes 1980; Gilmore 1987; Paz 1961:29–40; Robben 1988a; Stevens 1973).
7. Boat fishermen relay their own mistrust of female loyalty to the switching canoe fishermen by calling them cuckolds. Because of these insults, the switching canoe fishermen may become suspicious of their wives, may start an argument at home, and visit a brothel to prove their masculinity to their colleagues, their wives, and themselves.
8. The only exception was one brothel that had a couple of rooms in the back where a client could meet prostitutes who were sometimes left behind a locked door. The bar owner would inconspicuously hand the customer the key, he entered the room, and would lock it again when he left. This particular bar owner did act as a madam and would repeatedly contract new girls.

9. For an interesting urban contrast with the prostitution in small towns such as Camurim in which economic deprivation and stigmatization are the principal reasons for women to work as prostitutes, read Gaspar (1985). In her study of Copacabana prostitutes in Rio de Janeiro, she argues that the desire and sensation of social ascendency through dinners at expensive restaurants, invitations to luxurious yachts, and trips abroad are just as important for prostitutes as are the financial benefits.
10. Although McClelland's explanation of alcohol use is appropriate in this Brazilian case, it must not be attributed a universal validity as will become clear when I will discuss problem drinking in Camurium. Among the many other reasons mentioned for alcohol consumption in social science literature are: anxiety reduction, tension relief, escapism, weak or strong social norms of sobriety, overemphasis on self-reliance and achievement during childhood, improvement of social cohesion, prestige and status, adolescent rebellion, reaction against constrictive social rules, the sensation of personal strength and power, assertion of masculinity, relaxation, and spiritual insight (cf. Heath 1975; Mandelbaum 1979; Marshall 1979).
11. Pitt-Rivers (1977:9) has described the conflict between honor and legality very clearly: "For to go to law for redress is to confess publicly that you have been wronged and the demonstration of your vulnerability places your honour in jeopardy, a jeopardy from which the 'satisfaction' of legal compensation at the hands of a secular authority hardly redeems it. Moreover, it gives your offender the chance to humiliate you further by his attitude during all the delays of court procedure, which in fact can do nothing to restore your honour but merely advertises its plight."
12. The following references were used in my analysis of Brazilian patron-client relations: Forman and Riegelhaupt 1979; Freyre 1964; Galjart 1964; Greenfield 1972, 1979; Hutchinson 1966; Leal 1976; Robben 1984; Roniger 1987.
13. Canoe fishermen who switch to boat fishing and befriend boat owners may also be invited to weekend bars. This seems strange because it takes a long time before boat owners grant such favors to high-producing fishermen. The main reason is that boat owners want to retain these excellent fishermen for their crews. They know that these men will switch back to canoe fishing if their catches decline, hence they try to entice them with the promise of boat ownership. At the same time, these former canoe fishermen have—at least on the surface—adopted the conceptions of masculinity and indicators of prestige of the permanent boat fishermen. Despite past reservations about boat fishing, they are attracted by the prospect of greater wealth and status.

14. Alcoholism is either classified as a physical illness, or as an illness of the soul and the spirit. Medical doctors and herbalists treat the former; religious specialists cure the latter (Loyola 1982). Camurim does not have any medical facility to treat drinkers, but medicinal herbs are sold at the Sunday market. People who seek supernatural help pray in church, light candles, or visit local Candomblé centers (Leacock 1979).

Although most townspeople consider themselves members of the Roman Catholic Church, they nonetheless take the powers of Candomblé seriously. They believe that the Holy Trinity is more powerful than the *Orixás* (deities), but when prayers, vows, and candles do not cure a person, assistance will be sought at a Candomblé center. A man may be "addicted" (*viciado*) to alcohol because of a physical dependence, because of the evil eye, the spell of an envious colleague, or because he is dominated by a spirit or *encantado*. "It is believed that the individual's own spirit is forced outside his body by the encantado; it remains nearby, ready to re-enter when the encantado leaves" (Leacock and Leacock 1975:57). The leader of the Candomblé (*Mãe-do-Santo*) cures the man of the spell or allows the spirit to appear so that it can be controlled and exorcised. Most often, this spirit is the kingfisher bird (*Martim-Pescador*) who carries messages between the Gods and human beings.

After a fisherman goes into trance during a Candomblé session, *Martim-Pescador* takes control of his body and begins to act like a drunkard. He sways and staggers on his feet, pretends to drink cachaça from his right thumb, and talks incoherently. The women sing: "*Martim-Pescador,* what a life you have! Drinking *cachaça* and stumbling in the street" (cf. Carneiro 1978:75). The *Mãe-do-Santo,* who is also in trance, scolds *Martim-Pescador* and asks why he is bothering the fisherman and obliging him to drink so much while his wife and children suffer at home.

After several sessions, the fisherman may be able to control the spirit and his drinking. His recovery is hailed by relatives, friends, and neighbors who have once again reconciled themselves with him. The fisherman passes through a social rebirth in which he is reinstalled in his former position in the community. The period of alcohol abuse does not leave any blemish but is rather regarded as a valuable experience that made him into a more responsible person and showed his determination and willpower.

Several former alcoholics have become captains and have been elected to offices of the Fishermen's Guild that they stood little chance of occupying before their problem drinking began. In other words, although the people of Camurim believe that persons with weak personalities are more susceptible to problem drinking than are others, they recognize at the same time the strength of social and supernatural forces in educating a person to control his behavior. Furthermore, there exists forgiveness for the immoral behavior of former problem drinkers. See DaMatta (1987:167) for a discussion of the importance of the Catholic ethic of sin and salvation in Brazil.

15. For a detailed analysis of how the elite tries to hold on to its political power in Camurim, see Robben (1988b).
16. The term *o povão* (the masses) is also used by these men, but it usually refers to an emotionally charged mass of people—as in street riots or a delirious crowd of soccer fans.
17. The term *prestígio* has many meanings. In the context of vertical dyadic relationships, it refers to the reputation of a generous patron. Here, *prestígio* is best translated as rank, as a measure of a person's place in the social hierarchy. A bar, an object, a drink, or a location can all be objectivations of *prestígio* in the sense that those who control them occupy a high rank in the community.

## *9. Conclusion: Toward Interpretive Economic Anthropology*

1. I consciously avoid saying that they will *know* how to act in the correct way because I believe that experiential knowledge or practical wisdom of the right way to approach the surf cannot be made explicit. The explanations given by the fishermen serve only as rules of thumb that, at best, may guide them towards finding the right response in case of crisis. See the quotation from Michael Polanyi in chapter 1.
2. I have found the following secondary texts on Heidegger's conception of technology useful: Biemel (1976:92–113, 133–148); Hood (1972); Ihde (1983); Loscerbo (1981:129–153); Schirmacher (1983).
3. My article "Technical Innovation as Cultural Disclosure: Ecological Interpretation in Brazilian Waters" (in press) presents a detailed discussion of technology as disclosing nature and culture.
4. Heidegger and Marcuse believed that the hegemony of technology was so great that it tended to crowd out and overpower all other modes of disclosure. Despite the great contrast of their political convictions, they both saw a glimmer of hope in art as subverting the technological hegemony. Heidegger talks with yearning about ancient Greece where art and technology were described with the same term *techne*, and artists and craftsmen were called

*technites.* Art and technology were equally valued as modes of creation or disclosure. In the modern world, technology has become the dominant mode of disclosure, and only art might provide the liberating reflection that can break this hegemony (Heidegger 1977:314–317). Marcuse has taken a more Marxist perspective but also sees art as a way to attain freedom from our enslavement by a one-dimensional, technology society. "A work of art can be called revolutionary if, by virtue of the aesthetic transformation, it represents, in the exemplary fate of individuals, the prevailing unfreedom and the rebelling forces, thus breaking through the mystified (and petrified) social reality, and opening the horizon of change (liberation)" (Marcuse 1978:xi).

5. I use here an ancient Greek, non-Hegelian and non-Marxian conception of dialectics. The dialectic process, as the Greek philosophers understood it, is inevitably unrestricted in its direction because practice is defined as much by what it is as by what it is not. Practices are always in the flux of becoming, and what will be is ultimately subject to human volition. In the words of Perelman (1970:83; my translation): "Thus it would be a dialectic which would not necessarily lead towards a preexisting finality through a uniform and necessary development, but which would allow a certain room for human liberty with its possibilities of transcending every system, every given totality."
6. As June Nash (1979:6) writes in her study of Bolivian miners: "In the ideology of capitalist production, the moral issues of labor exploitation are considered extraneous to rational market relationships since labor is treated simply as another factor of production comparable to capital or rent. However, morality of necessity enters into assessment of the labor market, since the wage determines whether human life can be sustained and regenerated at a level corresponding with minimal social values."
7. See Hirschman (1977) and Myers (1983), for two excellent studies on how Western economic ideology—which refers to passion, desire, ambition, and especially self-interest as the causes of people's insatiable consumption needs—crystallized in the eighteenth century and found its culmination in Adam Smith's *The Wealth of Nations.*

# Bibliography

Abercrombie, Nicholas, Stephen Hill, and Bryan S. Turner. 1980. *The Dominant Ideology Thesis.* London: Allen and Unwin.

Acheson, James M. 1975. The Lobster Fiefs: Economic and Ecological Effects of Territoriality in the Maine Lobster Industry. *Human Ecology* 3(3):183–207.

—— 1981. Anthropology of Fishing. *Annual Review of Anthropology* 10:275–316.

—— 1985. The Social Organization of the Maine Lobster Market. In Stuart Plattner, ed., *Markets and Marketing*, pp. 105–130. Lanham, Md.: University Press of America.

Alexander, Paul. 1977. Sea Tenure in Southern Sri Lanka. *Ethnology* 16(3):231–251.

*Annaes do Archivo Público e Museu do Estado da Bahia.* 1923.

Arendt, Hannah. 1958. *The Human Condition.* Chicago: University of Chicago Press.

Ariès, Philippe. 1962. *Centuries of Childhood: A Social History of Family Life.* New York: Knopf.

Attewell, Paul. 1974. Ethnomethodology since Garfinkel. *Theory and Society* 1:179–210.

Aubert, Vilhelm. 1965. A Total Institution: The Ship. In *The Hidden Society*, pp. 236–258. Totowa, N.J.: Bedminster Press.

Azevedo, Thales de. 1965. Family, Marriage, and Divorce in Brazil. In Dwight B. Heath and Richard N. Adams, eds., *Contemporary Cultures and Societies of Latin America: A Reader in the Social Anthropology of Middle and South America and the Caribbean*, pp. 288–310. New York: Random House.

Bacelar, Jeferson Afonso. 1982. *A Família da Prostituta.* São Paulo: Editora Atica.

Bachelard, Gaston. 1969. *The Poetics of Space.* Boston: Beacon Press.

Bakhtin, M. M. 1981. *The Dialogic Imagination: Four Essays.* Austin: University of Texas Press.

Bandeira de Mello e Silva, Sylvio Carlos. 1967. A Divisão Regional da Bahia e Os Problemas de Equipamento. *Boletim Baiano de Geografia* 9:32–77.

Barreira e Castro, Clovis. 1981. Abrolhos: Sanctuário Ecológico na Costa da Bahia. *Geográfica Universal* 81:26–37.

Bastide, Roger. 1978. *The African Religions of Brazil: Toward a Sociology of the Interpenetration of Civilizations.* Baltimore: Johns Hopkins University Press.

Bataille, Georges. 1988. *The Accursed Share: An Essay on General Economy.* Vol. 1: *Consumption.* New York: Zone Books.

Berger, Peter and Hansfried Kellner. 1964. Marriage and the Construction of Reality. *Diogenes* 46:1–24.

Berger, Peter and Thomas Luckmann. 1966. *The Social Construction of Reality: A Treatise in the Sociology of Knowledge.* Garden City, N.Y.: Doubleday.

Biemel, Walter. 1976. *Martin Heidegger: An Illustrated Study.* New York: Harcourt Brace Jovanovich.

Blumer, Herbert. 1969. *Symbolic Interactionism: Perspective and Method.* Englewood Cliffs, N.J.: Prentice-Hall.

Borgmann, Albert. 1984. *Technology and the Character of Contemporary Life: A Philosophical Inquiry.* Chicago: University of Chicago Press.

Bourdieu, Pierre. 1973. The Berber House. In Mary Douglas, ed., *Rules and Meanings: The Anthropology of Everyday Knowledge,* pp. 98–110. Harmondsworth: Penguin Books.

—— 1977. *Outline of a Theory of Practice.* Cambridge: Cambridge University Press.

—— 1984. *Distinction: A Social Critique of the Judgement of Taste.* Cambridge: Harvard University Press.

—— 1987. *Choses Dites.* Paris: Éditions de Minuit.

Brandes, Stanley H. 1980. *Metaphors of Masculinity: Sex and Status in Andalusian Folklore.* Philadelphia: University of Pennsylvania Press.

Brandt, Andres von. 1972. *Fish Catching Methods of the World.* West Byfleet: Fishing News (Books).

Bricker, Victoria Reifler. 1976. Some Zinacanteco Joking Strategies. In Barbara Kirshenblatt-Gimblett, ed., *Speech Play,* pp. 51–62. Philadelphia: University of Pennsylvania Press.

Caldeira, Clovis. 1954. *Fazendas de Cacau na Bahia.* Rio de Janeiro: Ministério da Agricultura.

Carneiro, Edison. 1978. *Candomblés da Bahia.* 6th ed. Rio de Janeiro: Civilização Brasileira.

Cascudo, Luís da Câmara. 1954. *Dicionário do Folclore Brasileiro.* Rio de Janeiro: Ministério da Educação e Cultura.

*Censo Demográfico.* 1900, 1920, 1940, 1950, 1960, 1970, 1980. Fundação Instituto Brasileiro de Geografia e Estatística. Rio de Janeiro: IBGE.

Certeau, Michel de. 1984. *The Practice of Everyday Life.* Berkeley: University of California Press.

Chazan, Daniel, Juvenil Longo de Sousa, and Wilson Duarte de Almeida. 1977. *Arquitetura Contemporânea Brasileira: Criatividade e Inventividade.* São Paulo: Instituto Roberto Simonsen.

Cicourel, Aaron V. 1970. The Acquisition of Social Structure: Toward a Developmental Sociology of Language and Meaning. In Jack D. Douglas, ed. *Understanding Everyday Life,* pp. 136–168. Chicago: Aldine.

Clifford, James. 1983. On Ethnographic Authority. *Representations* 1(2):118–146.

Collomp, Alain. 1986. Familles: Habitations et cohabitations. In Philippe Ariès and Georges Duby, eds., *Histoire de la vie privée: De la Renaissance aux Lumières,* 3:500–541. Paris: Éditions du Seuil.

Cordell, John. 1978. Carrying Capacity Analysis of Fixed-Territorial Fishing. *Ethnology* 17(1):1–24.

—— 1989. Social Marginality and Sea Tenure in Bahia. In John Cordell, ed., *A Sea of Small Boats.* 210–230. Cambridge, Mass.: Cultural Survival.

Crapanzano, Vincent. 1980. *Tuhami: Portrait of a Moroccan.* Chicago: University of Chicago Press.

Dallmayr, Fred R. 1984. *Polis and Praxis: Exercises in Contemporary Political Theory.* Cambridge, Mass.: MIT Press.

Dalton, George and Jasper Köcke. 1983. The Work of the Polanyi Group: Past, Present and Future. In Sutti Ortiz, ed., *Economic Anthropology: Topics and Theories,* pp. 21–50. Lanham, Md.: University Press of America.

DaMatta, Roberto. 1981. *Carnavais, Malandros e Heróis: Para uma Sociologia do Dilema Brasileiro.* 3d. ed. Rio de Janerio: Zahar Editores.

—— 1982. As Raízes da Violência no Brasil: Reflexões de um Antropólogo Social. In Roberto DaMatta, Maria Célia Pinheiro Machado Paoli, et al., eds., *Violência Brasileira,* pp. 11–44. São Paulo: Editora Brasiliense.

—— 1986. *O que faz o brasil, Brasil?* 2d. ed. Rio de Janeiro: Editora Rocca Ltda.

—— 1987. *A Casa & a Rua: Espaço, Cidadania, Mulher e Morte no Brasil.* Rio de Janeiro: Editora Guanabara.

Davenport, William. 1960. Jamaican Fishing: A Game Theory Analysis. *Papers in Caribbean Anthropology* 58:3–11.

Dias, Manuel Nunes. 1974. Para a História do Cacau na Economia Atlântica. *Coleção Museu Paulista, Série de História* 2:7–74.

DiStasi, Lawrence. 1981. *Mal Occhio: The Underside of Vision.* San

Francisco: North Point Press.

Douglas, Mary. 1968. The Social Control of Cognition: Some Factors in Joke Perception. *Man* 3(3):361–376.

Dreyfus, Hubert L. 1980. Holism and Hermeneutics. *Review of Metaphysics* 34(1):3–23.

—— and Paul Rabinow. 1982. *Michel Foucault: Beyond Structuralism and Hermeneutics.* Chicago: University of Chicago Press.

Drummond, Didier. 1981. *Architectes des Favelas.* Paris: Bordas.

Duarte, Luiz Fernando Dias. 1981. Identidade Social e Padrões de "Agressividade Verbal" em um Grupo de Trabalhadores Urbanos. *Boletim do Museu Nacional* n.s. 36:1–33.

Dumont, Louis. 1977. *From Mandeville to Marx: The Genesis and Triumph of Economic Ideology.* Chicago: University of Chicago Press.

Dundes, Alan. 1983. Defining Identity through Folklore. In Anita Jacobson-Widding, ed., *Identity: Personal and Socio-Cultural*, pp. 235–261. *Acta Universitatis Upsaliensis, Uppsala Studies in Cultural Anthropology 5.*

Dundes, Alan, Jerry W. Leach, and Bora Özkök. 1972. The Strategy of Turkish Boys' Verbal Dueling Rhymes. In John J. Gumperz and Dell Hymes, eds., *Directions in Sociolinguistics: The Ethnography of Communication*, pp. 130–160. New York: Holt, Rinehart and Winston.

Dwyer, Kevin. 1979. The Dialogic of Ethnology. *Dialectical Anthropology* 4(3):205–224.

—— 1982. *Moroccan Dialogues: Anthropology in Question.* Baltimore: Johns Hopkins University Press.

Eckert, Ross D. 1979. *The Enclosure of Ocean Resources: Economics and the Law of the Sea.* Stanford, Calif.: Hoover Institution Press.

Elias, Norbert. 1978. *The Civilizing Process: The Development of Manners*, vol. 1. New York: Urizen Books.

Ellul, Jacques. 1964. *The Technological Society.* New York: Random House.

—— 1972. The Technological Order. In Carl Mitcham and Robert Mackey, eds., *Philosophy and Technology: Readings in the Philosophical Problems of Technology*, pp. 86–105. New York: Free Press.

Erikson, Erik H. 1959. *Identity and the Life Cycle.* New York: International Universities Press.

Fabian, Johannes. 1983. *Time and the Other: How Anthropology Makes Its Object.* New York: Columbia University Press.

Fernandez, James. 1974. The Mission of Metaphor in Expressive Culture. *Current Anthropology* 15(2):119–145.

Figueiredo, Mariza. 1983. The Socioeconomic Role of Women Heads of Family in a Brazilian Fishing Village. *Feminist Issues* 3(2):83–103.

Filho, Adonias. 1976. *Sul da Bahia: Chão de Cacau.* Rio de Janeiro: Editora Civilização Brasileira.

Forman, Shepard. 1970. *The Raft Fishermen: Tradition and Change in the Brazilian Peasant Economy*. Bloomington: Indiana University Press.

—— 1975. *The Brazilian Peasantry*. New York: Columbia University Press.

Forman, Shepard and Joyce F. Riegelhaupt. 1979. The Political Economy of Patron-Clientship: Brazil and Portugal Compared. In Maxine L. Margolis and William E. Carter, eds., *Brazil: Anthropological Perspectives*, pp. 379–400. New York: Columbia University Press.

Foucault, Michel. 1972. *The Archaeology of Knowledge and the Discourse on Language*. New York: Pantheon Books.

—— 1977. *Discipline and Punish: The Birth of the Prison*. New York: Random House.

—— 1978. *The History of Sexuality*. Vol. 1: *An Introduction*. New York: Pantheon Books.

—— 1982. The Subject and Power. In Hubert L. Dreyfus and Paul Rabinow, eds., *Michel Foucault: Beyond Structuralism and Hermeneutics*, pp. 208–226. Chicago: University of Chicago Press.

—— 1986. *The History of Sexuality*. Vol. 2: *The Use of Pleasure*. New York: Random House.

Fowler, Roger. 1985. Power. In Teun A. van Dijk, ed., *Handbook of Discourse Analysis*. Vol. 4: *Discourse Analysis in Society*, pp. 61–82. London: Academic Press.

Freyre, Gilberto. 1937. *Mucambos do Nordeste*. Rio de Janeiro: Ministério da Educação e Saúde.

—— 1961. *Sobrados e Mucambos: Decadência do Patriarcado Rural e Desenvolvimento do Urbano*. 3d. ed., vol. 1. Rio de Janeiro: Livraria José Olympio Editora.

—— 1964. The Patriarchal Basis of Brazilian Society. In Joseph Maier and Richard W. Weatherhead, eds., *Politics of Change in Latin America*, pp. 155–173. New York: Praeger.

Furtado, Celso. 1968. *The Economic Growth of Brazil: A Survey from Colonial to Modern Times*. Berkeley: University of California Press.

Gadamer, Hans-Georg. 1985. *Truth and Method*. New York: Crossroad.

Galjart, Benno. 1964. Class and "Following" in Rural Brazil. *América Latina* 3:3–23.

Garcez, Angelina Nobre Rolim and Antonio Fernando G. de Freitas. 1979. *Bahia Cacaueira: Um Estudo de História Recente*. Salvador: Centro Editorial e Didático da Universidade Federal da Bahia.

Garfinkel, Harold. 1967. *Studies in Ethnomethodology*. Englewood Cliffs, N.J.: Prentice-Hall.

Gaspar, Maria Dulce. 1985. *Garotas de Programa: Prostituição em Copacabana e Identidade Social*. Rio de Janeiro: Zahar Editor.

Giddens, Anthony. 1979. *Central Problems in Social Theory: Action, Structure and Contradiction in Social Analysis*. London: Macmillan.

—— 1983. Four Theses on Ideology. *Canadian Journal of Political and*

*Social Theory* 7(1–2):18–21.

—— 1984. *The Constitution of Society: Outline of a Theory of Structuration.* Cambridge: Polity Press.

Gilmore, David. 1977. The Social Organization of Space: Class, Cognition, and Residence in a Spanish Town. *American Ethnologist* 4(3):437–451.

—— 1987. *Aggression and Community: Paradoxes of Andalusian Culture.* New Haven, Conn.: Yale University Press.

Godelier, Maurice. 1978. The Object and Method in Economic Anthropology. In David Seddon, ed., *Relations of Production: Marxist Approaches to Economic Anthropology,* pp. 49–126. London: Frank Cass.

Goffman, Erving. 1961. *Asylums: Essays on the Social Situation of Mental Patients and other Inmates.* Garden City, N.Y.: Doubleday.

Gramsci, Antonio. 1971. *Selections from the Prison Notebooks.* New York: International Publisher.

Greenfield, Sidney M. 1972. Charwomen, Cesspools, and Road Building: An Examination of Patronage, Clientage, and Political Power in Southeastern Minas Gerais. In Arnold Strickon and Sidney M. Greenfield, eds., *Structure and Process in Latin America: Patronage, Clientage, and Power Systems,* pp. 71–100. Albuquerque: University of New Mexico Press.

—— 1979. Patron-Client Exchanges in Southeastern Minas Gerais. In Maxine L. Margolis and William E. Carter, eds., *Brazil: Anthropological Perspectives,* pp. 362–378. New York: Columbia University Press.

Gudeman, Stephen. 1978. *The Demise of a Rural Economy: From Subsistence to Capitalism in a Latin American Village.* London: Routledge and Kegan Paul.

—— 1986. *Economics as Culture: Models and Metaphors of Livelihood.* London: Routledge and Kegan Paul.

Guerrand, Roger-Henri. 1987. Espaces privés. In Philippe Ariès and Georges Duby, eds., *Histoire de la vie privée: De la Révolution à la Grande Guerre.* 4:324–411. Paris: Éditions du Seuil.

Hall, Edward T. 1969. *The Hidden Dimension.* Garden City, N.Y.: Doubleday.

Hardin, G. 1968. The Tragedy of the Commons. *Science* 162:1243–1248.

Hardoy, Jorge Enrique and Ana Maria Hardoy. 1978. The Plaza in Latin America: From Teotihuacán to Recife. *Cultures* 5(4):59–92.

Heath, Dwight B. 1975. A Critical Review of Ethnographic Studies of Alcohol Use. In R. J. Gibbins, Y. Israel, H. Kalant, R. E. Popham, W. Schmidt, and R. Smart, eds., *Research Advances in Alcohol and Drug Problems,* 2:1–92. New York: Wiley.

Heidegger, Martin. 1962. *Being and Time.* New York: Harper and Row.

—— 1975. The Origin of the Work of Art. In *Poetry, Language, Thought.* pp. 17–87. New York: Harper and Row.

—— 1977. The Question Concerning Technology. In *Basic Writings*. pp. 287–317. New York: Harper and Row.

—— 1982. *The Basic Problems of Phenomenology*. Bloomington: Indiana University Press.

Heye, Ana Margarete. 1980. A Questão da Moradia numa Favela do Rio de Janeiro ou Como ter *Anthropological Blues* sem Sair de Casa. In Gilberto Velho, ed., *O Desafio da Cidade: Novas Perspectivas da Antropologia Brasileira*, pp. 117–141. Rio de Janeiro: Editoria Campus.

Hirschkop, Ken. 1986. Bakhtin, Discourse and Democracy. *New Left Review* 160:92–113.

Hirschman, Albert O. 1977. *The Passions and the Interests: Political Arguments for Capitalism before its Triumph*. Princeton, N.J.: Princeton University Press.

Hood, Webster F. 1972. The Aristotelian Versus the Heideggerian Approach to the Problem of Technology. In Carl Mitchum and Robert Mackey, eds., *Philosophy and Technology: Readings in the Philosophical Problems of Technology*, pp. 347–363. New York: Free Press.

Howell, Richard W. 1973. Teasing Relationships. *Module in Anthropology* 46. Reading, Mass.: Addison-Wesley.

Hoy, David Couzens. 1978. *The Critical Circle: Literature, History, and Philosophical Hermeneutics*. Berkeley: University of California Press.

Hutchinson, Bertram. 1966. The Patron-Dependent Relationship in Brazil: A Preliminary Examination. *Sociologia Ruralis* 6(1):3–30.

Hymes, Dell. 1972. Models of the Interaction of Language and Social Life. In John J. Gumperz and Dell Hymes, eds. *Directions in Sociolinguistics: The Ethnography of Communication*, pp. 35–71. New York: Holt, Rinehart and Winston.

Ihde, Don. 1983. The Historical-Ontological Priority of Technology over Science. In Paul T. Durbin and Friedrich Rapp, eds., *Philosophy and Technology*, pp. 235–252. Dordrecht: D. Reidel.

Iutaka, Sugiyama. 1971. The Changing Bases of Social Class in Brazil. In John Saunders, ed., *Modern Brazil: New Patterns and Development*, pp. 257–268. Gainesville: University of Florida Press.

Keane, A. H. 1884. On the Botocudos. *Journal of the Anthropological Institute of Great Britain and Ireland* 13:199–213.

Keesing, Roger M. 1975. *Kin Groups and Social Structure*. New York: Holt, Rinehart and Winston.

Kottak, Conrad Phillip. 1983. *Assault on Paradise: Social Change in a Brazilian Village*. New York: Random House.

Kottak, Isabel Wagley. 1977. A Village Prostitute in Northeastern Brazil. *Michigan Discussions in Anthropology* 2:245–252.

Kuhn, Thomas S. 1962. *The Structure of Scientific Revolutions*. Chicago: University of Chicago Press.

Lavie, Smadar. 1984. The Fool and the Hippies: Ritual/Play and Social

Inconsistencies Among the Mzeina Bedouin of the Sinai. In Brian Sutton-Smith and Diana Kelly-Byrne, eds., *The Masks of Play*, pp. 63–70. New York: Leisure Press.

Leacock, Seth. 1979. Ceremonial Drinking in an Afro-Brazilian Cult. In Mac Marshall, ed., *Beliefs, Behaviors, and Alcoholic Beverages: A Cross-Cultural Survey*, pp. 81–93. Ann Arbor: University of Michigan Press.

—— and Ruth Leacock. 1975. *Spirits of the Deep: A Study of an Afro-Brazilian Cult.* Garden City, N.Y.: Doubleday.

Leal, Victor Nunes. 1976. *Coronelismo, Enxada e Voto: O Município e o Regime Representativo no Brasil.* São Paulo: Editora Alfa-Omega.

Lefebvre, Henri. 1971. *Everyday Life in the Modern World.* London: Penguin Press.

Leite, Serafim. 1938. *História da Companhia de Jesus no Brasil.* Tomo I. Lisboa: Livraria Portugália.

Leiter, Kenneth. 1980. *A Primer on Ethnomethodology.* New York: Oxford University Press.

Lewcock, Ronald and Gerald Brans. 1977. The Boat as an Architectural Symbol. In Paul Oliver, ed., *Shelter, Sign and Symbol*, pp. 107–116. Woodstock, N.Y.: Overlook Press.

Loscerbo, John. 1981. *Being and Technology: A Study in the Philosophy of Martin Heidegger.* The Hague: Martinus Nijhoff.

Loyola, Maria Andréa. 1982. Cure des corps et cure des âmes: Les rapports entre les médecines et les religions dans la banlieu de Rio. *Actes de la Recherche en Sciences Sociales* 43:3–45.

Luckmann, Thomas. 1983. Remarks on Personal Identity: Inner, Social and Historical Time. In Anita Jacobson-Widding, ed., *Identity: Personal and Socio-Cultural*, pp. 67–91. Acta Universitatis Upsaliensis, Uppsala Studies in Cultural Anthropology 5.

Magalhães de Gandavo, Pero de. 1922. *The Histories of Brazil.* vol. 2. New York: The Cortes Society.

Malinowski, Bronislaw. 1961. *Argonauts of the Western Pacific.* New York: E. P. Dutton.

Maloney, Clarence. 1976. Don't Say "Pretty Baby" Lest You Zap It with Your Eye—The Evil Eye in South Asia. In Clarence Maloney, ed., *The Evil Eye*, pp. 102–148. New York: Columbia University Press.

Mandelbaum, David G. 1979. Alcohol and Culture. In Mac Marshall, ed., *Beliefs, Behaviors, and Alcoholic Beverages: A Cross-Cultural Survey*, pp. 14–30. Ann Arbor: University of Michigan Press.

Marcuse, Herbert. 1964. *One-Dimensional Man: Studies in the Ideology of Advanced Industrial Society.* Boston: Beacon Press.

—— 1978. *The Aesthetic Dimension: Toward a Critique of Marxist Aesthetics.* Boston: Beacon Press.

Marshall, Mac. 1979. *Weekend Warriors: Alcohol in a Micronesian Culture.* Palo Alto, Cal.: Mayfield.

Matthiessen, Peter. 1975. *Far Tortuga.* New York: Random House.

McCay, Bonnie J. 1978. Systems Ecology, People Ecology, and the Anthropology of Fishing Communities. *Human Ecology* 6(4):397–422.

McClelland, David C. 1971. The Power of Positive Drinking. *New Society* 17(450):814–816.

—— 1984. Drinking as a Response to Power Needs in Men. In *Motives, Personality, and Society: Selected Papers,* pp. 327–342. New York: Praeger.

McClelland, David C., William N. Davis, Rudolf Kalin, and Eric Wanner. 1972. *The Drinking Man.* New York: Free Press.

Mead, George Herbert. 1977. *On Social Psychology.* Chicago: University of Chicago Press.

Merquior, J. G. 1985. *Foucault.* London: Fontana Press.

Miller, Charlotte I. 1979. The Function of Middle-Class Extended Family Networks in Brazilian Urban Society. In Maxine L. Margolis and William E. Carter, eds., *Brazil: Anthropological Perspectives,* pp. 305–316. New York: Columbia University Press.

Moore, Jr., Barrington. 1984. *Privacy: Studies in Social and Cultural History.* Armonk, N.Y.: M. E. Sharpe.

Moreira Pinto, Alfredo. 1899. *Apontamentos Para O Diccionario Geographico do Brazil.* Vol. 3. Rio de Janeiro: Imprensa Nacional.

Myers, Milton L. 1983. *The Soul of Modern Economic Man: Ideas of Self-Interest, Thomas Hobbes to Adam Smith.* Chicago: University of Chicago Press.

Nash, June. 1979. *We Eat the Mines and the Mines Eat Us: Dependency and Exploitation in Bolivian Tin Mines.* New York: Columbia University Press.

Nietschmann, Bernard. 1973. *Between Land and Water: The Subsistence Ecology of the Miskito Indians, Eastern Nicaragua.* New York: Seminar Press.

—— 1979. *Caribbean Edge: The Coming of Modern Times to Isolated People and Wildlife.* Indianapolis: Bobbs-Merrill.

Nóbrega, P. Manuel da. 1955. *Cartas do Brasil e Mais Escritos,* ed. Serafim Leite. Coimbra: Universidade de Coimbra.

Nolan, B. 1976. Seamen, Drink and Social Structure. *Maritime Policy and Management* 4(2):77–88.

Oliver, Paul. 1987. *Dwellings: The House across the World.* Austin: University of Texas Press.

Orlove, Benjamin S. 1986. Barter and Cash Sales on Lake Titicaca: A Test of Competing Approaches. *Current Anthropology* 27(2):85–106.

Ortiz, Sutti. 1983. What Is Decision Analysis About? The Problems of Formal Representations. In Sutti Ortiz, ed., *Economic Anthropology:*

*Topics and Theories*, pp. 249–297. Lanham, Md.: University Press of America.

Ortner, Sherry B. 1973. On Key Symbols. *American Anthropologist* 75(5):1338–1346.

—— 1984. Theory in Anthropology since the Sixties. *Comparative Studies in Society and History* 26(1):126–166.

Pálsson, Gísli and E. Paul Durrenberger. 1983. Icelandic Foremen and Skippers: The Structure and Evolution of a Folk Model. *American Ethnologist* 10(3):511–528.

Paul-Lévy, Françoise and Marion Segaud. 1983. *Anthropologie de l'espace.* Paris: Centre Georges Pompidou.

Paz, Octavio. 1961. *The Labyrinth of Solitude: Life and Thought in Mexico.* New York: Grove Press.

Pereira, Raul. 1976. *Peixes de Nossa Terra.* São Paulo: Livraria Nobel.

Perelman, Ch. 1970. Dialectique et Dialogue. In Rüdiger Bubner, ed., *Hermeneutik und Dialektik.* 2:77–84. Tübingen: J. C. B. Mohr.

Pierson, Donald. 1951. *Cruz das Almas: A Brazilian Village.* Washington, D.C.: Smithsonian Institution.

Pitt-Rivers, Julian. 1968. Honor. *International Encyclopedia of the Social Sciences.* New York: Macmillan & Free Press.

—— 1977. *The Fate of Shechem or The Politics of Sex: Essays in the Anthropology of the Mediterranean.* Cambridge: Cambridge University Press.

Plattner, Stuart. 1983. Economic Custom in a Competitive Marketplace. *American Anthropologist* 85(4):848–858.

Polanyi, Karl. 1957. The Economy as Instituted Process. In Karl Polanyi, Conrad M. Arensberg, and Harry W. Pearson, eds., *Trade and Market in the Early Empires*, pp. 243–270. Glencoe, Ill.: Free Press.

Polanyi, Michael. 1958. *Personal Knowledge: Towards a Post-Critical Philosophy.* London: Routledge and Kegan Paul.

Poole, Michael. 1981. Maritime Sociology: Towards a Delimitation of Themes and Analytical Frameworks. *Maritime Policy and Management* 8(4):207–222.

Popkin, Samuel. 1979. *The Rational Peasant: The Political Economy of Rural Society in Viet Nam.* Berkeley: University of California Press.

Poppino, Rollie E. 1973. *Brazil: The Land and People.* New York: Oxford University Press.

Rabinow, Paul. 1977. *Reflections on Fieldwork in Morocco.* Berkeley: University of California Press.

Radcliffe-Brown, A. R. 1952. *Structure and Function in Primitive Society.* London: Cohen and West.

Radcliffe-Brown, A. R. and Daryll Forde, eds. 1965. *African Systems of Kinship and Marriage.* London: Oxford University Press.

Ranum, Orest. 1986. Les Refuges de l'intimité. In Phillipe Ariès and

Georges Duby, eds. *Histoire de la vie privée: De la Renaissance aux Lumières.* 3:211–265. Paris: Éditions du Seuil.

*Revista do Instituto Geographico e Historico da Bahia.* 1902.

*Revista Trimensal do Instituto Geographico e Historico da Bahia.* 1895, 1896, 1898, 1899.

Ribeiro, Darcy. 1971. *The Americas and Civilization.* New York: E. P. Dutton.

Ribeiro, José. n.d. *Brasil no Folclore.* Rio de Janeiro: Gráfica Editora Aurora.

Ribeiro, René. 1945. On the *Amaziado* Relationship, and Other Aspects of the Family in Recife (Brazil). *American Sociological Review* 10(1): 44–51.

Richardson, Miles. 1982. Being-in-the-Market versus Being-in-the-Plaza: Material Culture and the Construction of Social Reality in Spanish America. *American Ethnologist* 9(2):421–436.

Ricoeur, Paul. 1974. *The Conflict of Interpretations: Essays in Hermeneutics.* Evanston, Ill.: Northwestern University Press.

Robben, Antonius C. G. M. 1984. Entrepreneurs and Scale: Interactional and Institutional Constraints on the Growth of Small-Scale Enterprises in Brazil. *Anthropological Quarterly* 57(3):125–138.

—— 1985. Sea Tenure and Conservation of Coral Reef Resources in Brazil. *Cultural Survival Quarterly* 9(1):45–47.

—— 1988a. Conflicting Gender Conceptions in a Pluriform Fishing Economy: A Hermeneutic Perspective on Conjugal Relationships in Brazil. In Jane H. Nadel-Klein and Dona Lee Davis, eds., *To Work and To Weep: Women in Fishing Economies,* pp. 106–129. St. John's, Newfoundland: Institute of Social and Economic Research, Memorial University.

—— 1988b. The Play of Power: Paradoxes in Brazilian Politics and Soccer. In Geert Banck and Kees Koonings, eds., *Social Change in Contemporary Brazil: Politics, Class and Culture in a Decade of Transition,* pp. 135–150. Dordrecht: Foris Publications.

—— In press. Technical Innovation as Cultural Disclosure: Ecological Interpretation in Brazilian Waters. In J. van den Breemer, H. Th. van der Pas, H. Thieleman, and W. Wolters, eds., *Innovation.* Leiden: Leidse Universiteitspers.

Roniger, Luis. 1987. Caciquismo and Coronelismo: Contextual Dimensions of Patron Brokerage in Mexico and Brazil. *Latin American Research Review* 22(2):71–100.

Rorty, Richard. 1979. *Philosophy and the Mirror of Nature.* Princeton: Princeton University Press.

Sahlins, Marshall D. 1976. *Culture and Practical Reason.* Chicago: University of Chicago Press.

—— 1979. *Stone Age Economics.* New York: Aldine.

—— 1981. *Historical Metaphors and Mythical Realities: Structure in the Early History of the Sandwich Islands Kingdom.* Ann Arbor: University of Michigan Press.

—— 1985. *Islands of History.* Chicago: University of Chicago Press.

Said, Edward W. 1983. *The World, the Text, and the Critic.* Cambridge: Harvard University Press.

Schirmacher, Wolfgang. 1983. From the Phenomenon to the Event of Technology: A Dialectical Approach to Heidegger's Phenomenology. In Paul T. Durbin and Friedrich Rapp, eds., *Philosophy and Technology,* pp. 275–289. Dordrecht: D. Reidel.

Schneider, David M. 1976. Notes toward a Theory of Culture. In Keith H. Basso and Henry A. Selby, eds., *Meaning in Anthropology,* pp. 197–220. Albuquerque: University of New Mexico Press.

Schrag, Calvin O. 1986. *Communicative Praxis and the Space of Subjectivity.* Bloomington: Indiana University Press.

Schutz, Alfred. 1976. The Homecomer. In *Collected Papers,* II: *Studies in Social Theory,* pp. 106–119. The Hague: Martinus Nijhoff.

Schwartz, Stuart B. 1973. *Sovereignty and Society in Colonial Brazil: The High Court of Bahia and Its Judges, 1609–1751.* Berkeley: University of California Press.

—— 1985. *Sugar Plantations in the Formation of Brazilian Society: Bahia, 1550–1835.* Cambridge: Cambridge University Press.

Scott, James C. 1976. *The Moral Economy of the Peasant: Rebellion and Subsistence in Southeast Asia.* New Haven: Yale University Press.

—— 1985. *Weapons of the Weak: Everyday Forms of Peasant Resistance.* New Haven: Yale University Press.

Sharrock, Wes and Bob Anderson. 1986. *The Ethnomethodologists.* New York: Tavistock Publications.

Sider, Gerald M. 1980. The Ties that Bind: Culture and Agriculture, Property and Propriety in the Newfoundland Village Fishery. *Social History* 5(1):1–39.

Soares de Sousa, Gabriel. 1940. *Notícia do Brasil.* tomo 1. ed. Pirajá da Silva. São Paulo: Livraria Martins Editora.

Souza, Luiz Carlos de. 1976. *Maralto: Relato de uma Pesca Perigosa.* Rio de Janeiro: Editora Civilização Brasileira.

Stallybrass, Peter and Allon White. 1986. *The Politics and Poetics of Transgression.* London: Methuen.

Stevens, Evelyn P. 1973. Machismo and Marianismo. *Society* 10(6): 57–63.

Taussig, Michael T. 1977. The Genesis of Capitalism Amongst a South American Peasantry: Devil's Labor and the Baptism of Money. *Comparative Studies in Society and History* 19(2):130–155.

—— 1980. *The Devil and Commodity Fetishism in South America.* Chapel Hill: University of North Carolina Press.

Taylor, Charles. 1979. Interpretation and the Sciences of Man. In Paul Rabinow and William M. Sullivan, eds., *Interpretive Social Science: A Reader,* pp. 25–71. Berkeley: University of California Press.

—— 1986. Foucault on Freedom and Truth. In David Couzens Hoy, ed., *Foucault: A Critical Reader,* pp. 69–102. Oxford: Basil Blackwell.

Terdiman, Richard. 1985. *Discourse/Counter-Discourse: The Theory and Practice of Symbolic Resistance in Nineteenth-Century France.* Ithaca, N.Y.: Cornell University Press.

Thomas, Georg. 1968. *Die portugiesische Indianerpolitik in Brasilien: 1500–1640.* Berlin: Colloquium Verlag Otto H. Hess.

Todorov, Tzvetan. 1984. *Mikhail Bakhtin: The Dialogical Principle.* Minneapolis: University of Minnesota Press.

Velho, Gilberto. 1975. *A Utopia Urbana: Um Estudo de Antropologia Social.* 2d ed. Rio de Janeiro: Zahar Editores.

Waterhouse, Roger. 1981. *A Heidegger Critique: A Critical Examination of the Existential Phenomenology of Martin Heidegger.* Brighton, Sussex: Harvester Press.

Watson, O. Michael. 1970. *Proxemic Behavior: A Cross-Cultural Study.* The Hague: Mouton.

Willems, Emilio. 1953. The Structure of the Brazilian Family. *Social Forces* 31(4):339–345.

Woortmann, Klaas. 1982. Casa e Família Operária. *Anuário Antropológico* 80:119–150.

Zulaika, Joseba. 1981. *Terranova: The Ethos and Luck of Deep-Sea Fishermen.* Philadelphia: Institute for the Study of Human Issues.

# Index